Heterogeneous Computing Archite~~ctures~~

Challenges a~~nd~~

Heterogeneous Computing Architectures
Challenges and Vision

Edited by
Olivier Terzo
Karim Djemame
Alberto Scionti
Clara Pezuela

CRC Press
Taylor & Francis Group
Boca Raton London New York

CRC Press is an imprint of the
Taylor & Francis Group, an **informa** business

CRC Press
Taylor & Francis Group
6000 Broken Sound Parkway NW, Suite 300
Boca Raton, FL 33487-2742

First issued in paperback 2022

© 2020 by Taylor & Francis Group, LLC
CRC Press is an imprint of Taylor & Francis Group, an Informa business

No claim to original U.S. Government works

ISBN-13: 978-0-367-02344-7 (hbk)
ISBN-13: 978-1-03-233804-0 (pbk)
DOI: 10.1201/9780429399602

Library of Congress Cataloging-in-Publication Data

Names: Terzo, Olivier, editor.
Title: Heterogeneous computing architectures : challenges and vision /
Olivier Terzo, Karim Djemame, Alberto Scionti and Clara Pezuela.
Description: Boca Raton : Taylor & Francis, a CRC title, part of the Taylor
& Francis imprint, a member of the Taylor & Francis Group, the
academic division of T&F Informa, plc, 2019. | Includes bibliographical
references and index.
Identifiers: LCCN 2019021391 | ISBN 9780367023447 (hardback : acid-free
paper) | ISBN 9780429399602 (e-book)
Subjects: LCSH: Heterogeneous computing.
Classification: LCC QA76.88 .T49 2019 | DDC 004/.35--dc23
LC record available at https://lccn.loc.gov/2019021391

Visit the Taylor & Francis Web site at
http://www.taylorandfrancis.com

and the CRC Press Web site at
http://www.crcpress.com

Contents

4   Simplifying Parallel Programming and Execution for Distributed Heterogeneous Computing Platforms   89
J. Ejarque, D. Jiménez-González, C. Álvarez, X. Martorell, and R. M. Badia

5   Design-Time Tooling to Guide Programming for Embedded Heterogeneous Hardware Platforms   103
R. De Landtsheer, J. C. Deprez, and L. Guedria

**9   Machine Learning on Low-Power Low-Cost Platforms: An
     Application Case Study   191**
     A. Scionti, O. Terzo, C. D'Amico, B. Montrucchio, and R. Ferrero

10 Security for Heterogeneous Systems   221
     M. Tsantekidis, M. Hamad, V. Prevelakis, and M. R. Agha

Foreword

I am very pleased that the Heterogeneity Alliance decided to publish a book on the challenges and visions of heterogeneous computing architectures; it is both important and timely. With the end of Dennard scaling, the "free lunch" of automatic performance scaling with every processor generation has ended. Multi-cores and many-cores were deemed the best use of the growing number of transistors. This transition however put the burden of performance scaling on the software developers because suddenly performance scaling had to come from scaling up the application parallelism. Now that Moore's law is also slowing down, and will eventually end, the number of cores per chip will also stop growing, and so will this type of performance scaling.

The only known solution to further scale performance is to customize the hardware for particular application domains. Advanced specialization can lead to performance improvements of up to three orders of magnitude compared to general-purpose processors, so there is still a lot of potential. Popular examples of accelerators are GPUs, TPUs, DSPs, and FPGAs. Unfortunately, accelerators do not come free: designing them is expensive, and programming them requires advanced specialty knowledge. Furthermore, bringing a new accelerator to the market is a challenge: besides the hardware, there is a need for advanced tools like compilers, profilers, and a run-time system; the hardware has to be integrated into existing platforms and it has to amass an active user base, i.e., it needs to grow a complete ecosystem.

In the past, many accelerators were caught up after a few years by general-purpose processors, and it was difficult to find a solid business case for a new accelerator. This time is different: there is almost a guarantee that the accelerator will stay ahead of a general-purpose core for many years. Future advanced computing platforms will contain several accelerators. This creates formidable challenges: What will be the new abstractions that hide the complexity of the hardware platform? How to develop tools and applications for such complex platforms? How to guarantee safety and security? How to secure programmer productivity? etc. This book deals with all these challenges.

This evolution toward more heterogeneity might eventually lead us into a new era in which software development becomes too complex for humans, and will have to be taken over by programs. This will not only be a solution for the complexity and productivity challenges, but also for the global lack of ICT workers. This evolution will open up a completely new sub-domain in software engineering. But for all this to happen, more groundbreaking research in software engineering is needed. I, therefore, encourage the reader to consider engaging in this endeavor.

Koen De Bosschere
Ghent, 16 December 2018

Acknowledgments

We would like to express our gratitude to all the professors and researchers who contributed to this book and to all those who provided support, talked things over, read, wrote, and offered comments.

We thank all authors and their organizations that allowed sharing relevant studies of scientific applications in the field of heterogeneous computing architectures.

A special thanks goes to the HiPEAC (European Network Excellence in High Performance and Embedded Architecture and Compilation) initiative supporting European researchers and industry representatives in computing systems; in particular, we want to thank Koen De Bosschere (coordinator) and Marc Duranton from HiPEAC.

We wish to thank our institutions, LINKS Foundation – Leading Innovation and Knowledge for Society, University of Leeds, Atos, which enabled us to research the fields relevant to the writing of this book. Dr. A. Scionti and Dr. O. Terzo want to thank their president Dr. Prof. Marco Mezzalama, their director Dr. Massimo Marcarini, and their colleagues from the ACA research unit for their precious suggestions and comments.

A special thanks to our publisher, Nora Konopka, for allowing this book to be published and all persons from Taylor & Francis Group who provided help and support at each step of the writing process.

We want to offer a sincere thank you to all the readers and all the persons who will promote this book.

About the Editors

Olivier Terzo holds a PhD in electronic engineering and communications and an MSc in computer engineering from Politecnico di Torino, Italy. He also holds a university degree in electrical engineering technology and industrial informatics from University Institute of Nancy, France.

Since 2013, Dr. Terzo has been head of the Advanced Computing and Applications (ACA) at the Leading Innovation and Knowledge for Society (LINKS) Foundation, which focuses on study, design, and implementation of computing infrastructures based on Cloud computing technologies. The main fields of research for the area are on research and application orchestration for distributed infrastructures, machine learning (ML) techniques and their applications, development of ML and other applications on FPGA accelerators, Cloud and ultra-low power systems convergence, and, in general, low power computing and communication.

Dr. Terzo was technical and scientific coordinator of the H2020 OPERA project. Currently, he is scientific and technical coordinator of the H2020 LEXIS project and of the Plastic & Rubber Regional project, which aims to respond to the Industry 4.0 demands. From 2010–2013 he was head of the Infrastructure Systems for Advanced Computing (IS4AC) Research Unit at LINKS Foundation, whose activities were focused on Cloud computing systems and infrastructures. Previously, from 2004–2009, he was researcher in the e-security laboratory at LINKS Foundation, focusing on P2P protocols, encryption on embedded devices, security of routing protocols, and activities on grid computing infrastructures. He is an active member of the HiPEAC community, where he co-organizes the workshop on Heterogeneous and Low Power Data Center Technologies (HeLP-DC). He is also active within the Heterogeneous Hardware & Software Alliance (HH&S) and the ETP4HPC community. He co-organizes international workshops, and he is associate editor member of the *International Journal of Grid and Utility Computing (IJGUC)*. Dr. Terzo is also an IPC member of the International Workshop on Scalable Optimization in Intelligent Networking, as well as a peer reviewer for ICNS and CISIS international conferences.

Karim Djemame is a professor at the School of Computing at the University of Leeds, UK, and is the co-founder of the Heterogeneous Hardware & Software Alliance (HH&S), an initiative undertaken by the Transparent Heterogeneous Hardware Architecture Deployment for the eNergy Gain in Operation (TANGO) project, which aims to join efforts of organizations interested in the development of future technologies and tools to advance and take full advantage of computing and applications using heterogeneous hardware.

He develops fundamental principles and practical methods for problems that are computationally challenging and/or require unusual kinds of computing resources. In particular, he focuses on HPC/Cloud/grid-based services to enable organizations to more efficiently manage their high-end computing resources. He previously led many UK research projects including the Leeds element of the e-Science DAME project with industrial partners Rolls Royce and Cybula. He has extensive experience in the research and management of EU projects (AssessGrid, OPTIMIS, ASCETiC, TANGO) and sits on a number of international programme committees for grid/Cloud middleware, computer networks, and performance evaluation, and is the Scientific and Technical Manager of TANGO.

Prof. Djemame's main research area is distributed systems, including system architectures, edge/Cloud computing, energy efficiency, resource management, and performance evaluation. He has published 160+ papers in magazines, conferences, and journals including *IEEE Transactions Series*. He is a member of the IEEE and the British Computer Society.

Alberto Scionti holds an MSc and a PhD (European doctorate degree) in computer and control engineering from Politecnico di Torino, Italy.

Currently, he is a Senior Researcher in the Advanced Computing and Applications (ACA) Research Area at Leading Innovation & Knowledge for Society (LINKS) Foundation. His main research areas comprise hardware design and testing of digital circuits (especially embedded memories), design and simulation of modern computer architectures, and high-performance computing systems (HPC and HPEC) spanning traditional HPC-like infrastructures to Cloud-based architectures to hybrid systems. He has much experience in designing hardware schedulers for data-flow CMPs, as well as design and simulation of networks-on-chip supporting data-flow and fine-grain threading execution models. Past experiences include also the application of evolutionary algorithms to industrial problems (e.g., generation of test patterns for digital circuits and drift removal from electronic nose systems). During his research activity, he amassed a strong knowledge and experience of lightweight virtualization systems (e.g., Docker, LXC/LXD, etc.), their applications in the context of distributed infrastructures, as well as exploitation of hardware accelerators for application-performance improvement. During his research and investigation activity, he cultivated expertise and knowledge in using high-level synthesis (e.g., Intel OpenCL) for porting deep learning algorithms on FPGA architectures, as well as to support the acceleration of modern computational heavy algorithms.

He was involved in several national and EU-funded projects (TERAFLUX, ERA, Demogrape, OPERA). Currently, he is involved in the EU-H2020 LEXIS project. He is co-author of more than 40 publications on peer-reviewed international journals, workshops, conferences, and book chapters, and peer reviewer for several international conferences, workshops, and journals. He is an active member of the HiPEAC (the European Network of Excellence for High Performance and Embedded Architecture and Compilation) and he is involved in the activities of the Heterogeneous Hardware & Software Alliance (HH&S) and ETP4HPC initiatives. Within the HiPEAC events, he co-organizes the Workshop on Heterogeneous and Low Power Data Center Technologies (HeLP-DC).

Clara Pezuela has a degree in computer science from the Universidad Politécnica of Madrid. She has 19 years' experience in R&D projects development and management. Currently, she is the head of IT Market at Research and Innovation Group in Atos. Her main responsibilities are the management of research groups focused on Cloud, digital platforms, intelligent systems, and software engineering, dealing with management of teams and projects, defining the strategy toward future research in mentioned topics and transferring research assets to Atos business units. She is skilled in open business models and innovation processes, collaborative development environments, and service and software engineering.

In the past years, she has coordinated an integrated project in FP7-ICT-ARTIST about migration of applications to the Cloud and an activity project (MCloudDaaS) in EIT Digital on usage of multi-Cloud in Big Data analytics as a service. Currently, she is coordinating a H2020-ICT-TANGO on simplifying and optimizing software development for heterogeneous devices and OFION, an EIT Digital project on orchestration of financial services. She is the FIWARE solution manager at Atos, an open and standard platform for development of smart applications in IoT. She is also the president of PLANETIC, the Spanish technology platform for the adoption and promotion of ICT in Spain. Her current interest areas are innovation management, the improvement of software development processes and methods, and the adoption of innovation assets by the industry.

Contributors

Mustafa R. Agha
Institute of Computer and Network
   Engineering – Technical
   University of Braunschweig
Braunschweig, Germany

Orestis Akrivopoulos
Spark Works ITC
Derbyshire, United Kingdom

Carlos Álvarez
Barcelona Supercomputing Center
   (BSC), Universitat Politècnica de
   Catalunya (UPC)
Barcelona, Spain

Rosa M. Badia
Barcelona Supercomputing Center
   (BSC), Spanish National Research
   Council (CSIC)
Barcelona, Spain

Konstantin Bakanov
Telecommunication Systems
   Institute (TSI), Technical
   University of Crete
Crete, Greece

Constantinos Bitsakos
National Technical University of
   Athens
Athens, Greece

Stefan Bonfert
Ulm University
Ulm, Germany

Paolo Burgio
Università degli Studi di Modena e
   Reggio Emilia
Modena, Italy

Roberto Cavicchioli
Università degli Studi di Modena e
   Reggio Emilia
Modena, Italy

Carmine D'Amico
LINKS Foundation – Leading
   Innovation & Knowledge for
   Society
Torino, Italy

Dimitrios Danopoulos
National Technical University of
   Athens (NTUA)
Athens, Greece

Renaud De Landtsheer
Centre d'Excellence en Technologies
   de l'Information et de la
   Communication (CETIC)
Gosselies, Belgium

Jean-Christophe Deprez
Centre d'Excellence en Technologies
   de l'Information et de la
   Communication (CETIC)
Gosselies, Belgium

Katerina Doka
National Technical University of
   Athens
Athens, Greece

Jorge Ejarque
Barcelona Supercomputing Center
  (BSC)
Barcelona, Spain

Renato Ferrero
Politecnico di Torino
Torino, Italy

Juan Alfonso Fumero
The University of Manchester
Manchester, United Kingdom

Konstantinos Georgopoulos
Telecommunication Systems
  Institute (TSI), Technical
  University of Crete
Crete, Greece

Lotfi Guedria
Centre d'Excellence en Technologies
  de l'Information et de la
  Communication (CETIC)
Gosselies, Belgium

Mohammad Hamad
Institute of Computer and Network
  Engineering – Technical
  University of Braunschweig
Braunschweig, Germany

Aggelos Ioannou
Technical University of Crete
Crete, Greece

Daniel Jiménez-González
Barcelona Supercomputing Center
  (BSC), Universitat Politècnica de
  Catalunya (UPC)
Barcelona, Spain

Christoforos Kachris
Institute of Communication and
  Computer Systems (ICCS)
Athens, Greece

Nikos Kanakis
Spark Works ITC
Derbyshire, United Kingdom

Richard Kavanagh
University of Leeds
Leeds, United Kingdom

Dirk Koch
The University of Manchester
Manchester, United Kingdom

Ioannis Konstantinou
National Technical University of
  Athens
Athens, Greece

Christos-Efthymios Kotselidis
The University of Manchester
Manchester, United Kingdom

Luciano Lavagno
Politecnico di Torino
Torino, Italy

Francesco Lubrano
LINKS Foundation – Leading
  Innovation & Knowledge for
  Society
Torino, Italy

Pavlos Malakonakis
Technical University of Crete
Crete, Greece

Xavier Martorell
Barcelona Supercomputing Center
  (BSC), Universitat Politècnica de
  Catalunya (UPC)
Barcelona, Spain

Iakovos Mavroidis
Telecommunication Systems
  Institute (TSI), Technical
  University of Crete
Crete, Greece

Somnath Mazumdar
Simula Research Laboratory
Lysaker, Norway

Bartolomeo Montrucchio
Politecnico di Torino
Torino, Italy

Ioannis Mytilinis
National Technical University of
Athens
Athens, Greece

Michail Papadimitriou
The University of Manchester
Manchester, United Kingdom

Ioannis Papaefstathiou
Telecommunication Systems
Institute (TSI), Technical
University of Crete
Crete, Greece

Khoa Pham
The University of Manchester
Manchester, United Kingdom

Vassilis Prevelakis
Institute of Computer and Network
Engineering – Technical
University of Braunschweig
Braunschweig, Germany

Lutz Schubert
Ulm University
Ulm, Germany

Dimitrios Soudris
National Technical University of
Athens (NTUA), Institute of
Communication and Computer
Systems (ICCS)
Athens, Greece

Marinos Tsantekidis
Institute of Computer and Network
Engineering – Technical
University of Braunschweig
Braunschweig, Germany

Christos Tselios
Spark Works ITC
Derbyshire, United Kingdom

Micaela Verucchi
Università degli Studi di Modena e
Reggio Emilia
Modena, Italy

Stefan Wesner
Ulm University
Ulm, Germany

Foivos Zakkak
The University of Manchester
Manchester, United Kingdom

1

Heterogeneous Data Center Architectures: Software and Hardware Integration and Orchestration Aspects

A. Scionti, F. Lubrano, and O. Terzo

LINKS Foundation – Leading Innovation & Knowledge for Society, Torino, Italy

S. Mazumdar

Simula Research Laboratory, Lysaker, Norway

CONTENTS

Machine learning (ML) and deep learning (DL) algorithms are emerging as the new driving force for the computer architecture evolution. With an ever-larger adoption of ML/DL techniques in Cloud and high-performance computing (HPC) domains, several new architectures (spanning from chips to entire

1

systems) have been pushed on the market to better support applications based on ML/DL algorithms. While HPC and Cloud remained for long time distinguished domains with their own challenges (i.e., HPC looks at maximising FLOPS, while Cloud at adopting COTS components), an ever-larger number of new applications is pushing for their rapid convergence. In this context, many accelerators (GP-GPUs, FPGAs) and customised ASICs (e.g., Google TPUs, Intel Neural Network Processor—NNP) with dedicated functionalities have been proposed, further enlarging the data center heterogeneity landscape. Also Internet of Things (IoT) devices started integrating specific acceleration functions, still aimed at preserving energy. Application acceleration is common also outside ML/DL applications; here, scientific applications popularised the use of GP-GPUs, as well as other architectures, such as Accelerated Processing units (APUs), Digital Signal Processors (DSPs), and many-cores (e.g., Intel XeonPhi). On one hand, training complex deep learning models requires powerful architectures capable of crunching large number of operations per second and limiting power consumption; on the other hand, flexibility in supporting the execution of a broader range of applications (HPC domain requires the support for double-precision floating-point arithmetic) is still mandatory. From this viewpoint, heterogeneity is also pushed down: chip architectures sport a mix of general-purpose cores and dedicated accelerating functions.

Supporting such large (at scale) heterogeneity demands for an adequate software environment able to maximise productivity and to extract maximum performance from the underlying hardware. Such challenge is addressed when one looks at single platforms (e.g., Nvidia provides CUDA programming framework for supporting a flexible programming environment, OpenCL has been proposed as a vendor independent solution targeting different devices— from GPUs to FPGAs); however, moving at scale, effectively exploiting heterogeneity remains a challenge. Whenever large number of heterogeneous resources (with such variety) must be managed, it poses new challenges also. Most of the tools and frameworks (e.g., OpenStack) for managing the allocation of resources to process jobs still provide a limited support to heterogeneous hardware.

The chapter contribution is two fold: *i*) presenting a comprehensive vision on hardware and software heterogeneity, covering the whole spectrum of a modern Cloud/HPC system architecture; *ii*) presenting ECRAE, i.e., an orchestration solution devised to explicitly deal with heterogeneous devices.

1.1 Heterogeneous Computing Architectures: Challenges and Vision

Cloud computing represents a well-established paradigm in the modern computing domain showing an ever-growing adoption over the years. The continuous improvement in processing, storage and interconnection technologies, along with the recent explosion of machine learning (ML) and more specifically of deep learning (DL) technologies has pushed Cloud infrastructures beyond their traditional role. Besides traditional services, we are witnessing at the diffusion of (Cloud) platforms supporting the ingestion and processing of massive data sets coming from ever-larger sensors networks. Also, the availability of a huge amount of cheap computing resources attracted the scientific community in using Cloud computing resources to run complex HPC-oriented applications without requiring access to expensive supercomputers [175, 176]. Figure 1.1 shows the various levels of the computing continuum, within which heterogeneity manifest itself.

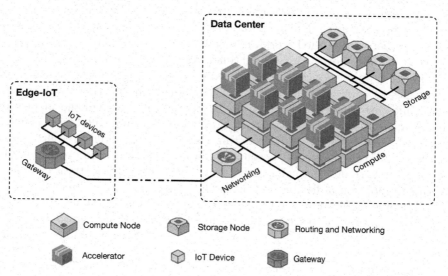

FIGURE 1.1: The heterogeneity over the computing continuum: From the IoT level (smart-connected sensors) to the data center (compute nodes and accelerators).

In this context data centers have a key role in providing necessary IT resources to a broad range of end-users. Thus, unlike past years, they became more heterogeneous, starting to include processing technologies, as well as storage and interconnection systems that were prerogative of only high-performance systems (i.e., high-end clusters, supercomputers). Cloud providers have been pushed to embed several types of accelerators, ranging from well known (GP-)GPUs to reconfigurable devices (mainly FPGAs) to more exotic hardware (e.g., Google TPU, Intel Neural Network Processor—NNP). Also, CPUs became more variegated in terms of micro-architectural features and ISAs. With ever-more complex chip-manufacturing processes, heterogeneity reached IoT devices too: here, hardware specialisation minimises energy consumption. Such vast diversity in hardware systems, along with the demand for more energy efficiency at any level and the growing complexity of applications, make mandatory rethinking the approach used to manage Cloud resources.

Among the others, recently, FPGAs emerged as a valuable candidate providing an adequate level of performance, energy efficiency and programmability for a large variety of applications. Thanks to the optimal trade-off between performance and flexibility (high-level synthesis—HLS—compilers ease mapping between high-level code and circuit synthesis), most of the (scientific) applications nowadays benefit from FPGA acceleration. Energy (power) consumption is the main concern for Cloud providers, since the rapid growth of the demand for computational power has led to the creation of large-scale data centers and consequently to the consumption of enormous amounts of electricity power resulting in high operational costs and carbon dioxide emissions. On the other hand, in IoT world, reducing energy consumption is mandatory to keep working devices for longer time, even in case of disconnection from the power supply grid. Data centers need proper (energy) power management strategies to cope with the growing energy consumption, that involves high economic cost and environmental issues. New challenges emerge in providing adequate management strategies. Specifically, resource allocation is made more effective by dynamically selecting the most appropriate resource for a given task to run, taking into account availability of specialised processing elements. Furthermore, through a global view of the data center status combined with the adoption of flexible and lightweight virtualization technologies, it is also possible to optimise the whole workload allocation. However, such optimisation must take into account several factors including the type of hardware resources available and its related constraints, the type of incoming workload and the constraints deriving from the SLA and QoS.

This chapter can be divided into two different parts: in the first part it will present a comprehensive set of state-of-the-art heterogeneous systems, which are becoming common architectural elements of modern data centers, but also in highly integrated devices (IoT); the second will introduce the resource allocation problem in the perspective of heterogeneous resources, and describe a corresponding orchestration solution, as devised within the European H2020 project OPERA [338]. The capabilities of the whole proposed solution are also demonstrated, through simulations and experiments on a test-bed using real workloads.

1.2 Backgrounds

The Cloud computing (CC) model was first introduced by Eric Schmidt (Google Inc. USA) in 2006 [97]. According to the U.S. National Institute of Standards and Technology (NIST), CC provides inexpensive, ubiquitous, on-demand access to a shared pool of configurable computing resources [273].

A large fraction of these resources are located on Cloud provider data centers. However, with the emergence of applications relying on the IoT paradigm and demanding (near) real-time processing, CC started embracing also edge computing resources [343]. On the other hand, the IoT paradigm enables the massive generation and ingestion of data, that requires high computing capacities to be analysed. This demand, along with the energy-efficiency concern, forced the semiconductor industry to bring new kinds of processors on the market. More and more functions are added in every new processor generation, while limits in the actual usage of available transistors are emerging [145][146]. If until 2005 it was still possible to increase the clock frequency, limitation in the way heat may be dissipated put an end to the commercial race of higher clocked microprocessors. To overcome this issue, every new processor generation integrates more cores than previous ones. Nowadays, *multi-/many-core* architectures are available as commodity hardware. Nonetheless, architectural specialisation is becoming important in order to improve performance and power efficiency. In modern processors the on-chip storage capacity improved as well: old processors, with a single level of off-chip cache memory, have been substituted by newer chips equipped with multi-level on-chip cache memory systems. Similarly, simple bus interconnections have been substituted by more efficient crossbar switches, rings, and networks-on-chip (NoCs). Such kind of interconnections also made it possible to integrate modules that previously needed an external piece of silicon, without negatively impacting the processor performance. According to hardware changes (i.e., hardware become more specialised), also programming models have to adapt to the new available functionalities, in such way programmers can easily exploit them. By leveraging programming models such as OpenMP, MPI, MapReduce, CUDA or OpenCL, high-performance system designers and Cloud providers made their applications easily access heterogeneous hardware resources (the concept was first conceived by Chandrakasan et. al. in his paper [99]), such as Graphical Processing Units (GPUs), Digital Signal Processors (DSPs), customized hardware implemented on top of Field-Programmable Gate Arrays (FPGAs), and many others.

Modern data centers have made huge progress from the computing, networking, and virtualization points of view. The limits reached by the evolution of general-purpose processors (GPPs), with the consequent limitations to the constant increase of CPU performance defined in the Moore's law, have led data center providers to insert in their facilities specialized hardware, such as application's accelerators. The introduction of heterogeneous hardware in data centers raised workload management issues, due to the reduction of the homogeneity in the data center structure. For instance, in [94], a new and different architecture for the data center is presented: FPGAs, that commonly are an embedded component of the server, are placed between the server and the switch, thus allowing FPGAs to communicate directly through the network. A detailed survey on FPGAs in Cloud data centers are presented in [85].

The target of the solution devised in the EU project OPERA is to use accelerators to reach better performance and to reduce energy consumption in data centers. The energy consumption in data centers of all sizes is a problem that has been analyzed and addressed many times and consequently several theories have been proposed. The overall consumption consists of the sum of the consumption of its macro elements. The cooling system can constitute up to 50% of the energy consumption, while IT equipment constitutes 36%. The remainder of the power consumption is mainly ascribed to (inherent) imperfection of the power conversion process [123]. Regarding the IT facilities, techniques such as Dynamic Voltage and Frequency Scaling (DVFS) allow reduced energy consumption acting at the hardware level (e.g., reducing the clock frequency of components), but maintains the performances unaltered. Better reduction of power consumption can be achieved through better resource provisioning. In [66] are presented the results of several simulations that show how it is possible to reduce the energy consumption by applying an allocation policy which aims to allocate VMs in the most efficient hosts compared to a Best Fit allocation policy. After a first VMs allocation, a consolidation process decides if migrating VMs is in order to switch off idle nodes. This technique allows a significant reduction of energy consumption in front of an overhead introduced by the migration process.

In [167] is introduced the concept of *hybrid provisioning*. This type of resource provisioning consists of a two-times approach: i) the VM workload is predicted and brought back to a model, then resources are allocated based on the prediction; ii) at run-time it evaluates the deviation between the real workload and the predicted one and whether the VMs are migrated on better hosts.

1.2.1 Microprocessors and Accelerators Organization

A modern microprocessor (and also an accelerator) contains several billions of transistors, and several levels of metallization to connect them to each other. Such impressive numbers of components are organized as computing cores, interconnections, and memory blocks. This complexity requires that many aspects are taken into account when such systems are analyzed (e.g., manufacturing technology, interconnections, internal micro-architecture, etc.).

Manufacturing process. Although several times the scientific community has announced the end of the Moore's law, technology has always been able to increase the number of integrated transistors at every new generation [210]. Nevertheless, complementary metal-oxyde-semiconductor (CMOS) technology became predominant in the last two decades. Current technology, still based on CMOS technology, uses multi-gate (e.g., FinFET transistors) and silicon-on-insulator (SOI) transistors [109][228]. Recently, 3D chip stacking has been proposed as a way to increase the density of computing elements per package, by connecting more than one silicon die [255], however, their appearance on the market is still far from now.

Micro-architecture organisation. Micro-architectural aspects concern the organisation of transistors into functional units. Grouping these functional units into a suitable form allows the implementation of computing cores ranging from simple not pipelined in-order micro-architectures to out-of-order superscalar ones. The simplest micro-architecture is represented by a processing element that processes up to one instruction per cycle. In this case, the instructions are executed in the order specified by the programmer (i.e., *in-order architectures*). Since several operations can be involved in the execution of a single instruction, *pipelined* processors try to improve the number of committed instructions per cycle by overlapping these operations among subsequent instructions. Although, they can reach an ideal IPC value, temporal dependency among subsequent instructions can negatively affect their execution. With the aim of overcoming those limitations, *out-of-order–OoO* micro-architectures execute the instruction stream in an order that is different from that expressed by the programmer, thus obtaining an IPC value greater than 1. However, out-of-order processors require a complex hardware scheduler that is responsible for issuing the instructions only depending on the availability of input data, coupled with a re-order unit that commits instructions in the order specified by the programmer. A different approach is based on static execution of the code that is arranged by the compiler in such a way multiple independent instructions are packet together. Micro-architectures that follow this design approach are called *very long instruction word* (VLIW). Over time, several approaches have been explored by designers to improve the processor performance. Higher clock speed can be used, by adopting very deep pipelines (i.e., generally, these processors adopt pipelines with more than 20 stages). Alternatively, designers can increase the number of execution pipelines, so that more than one instruction can be executed in parallel. Processors that follow this philosophy are defined as *superscalar*. *Vector processors* have been successfully used in past supercomputers (e.g., the series of past Cray supercomputers). Nowadays, vector processors are embedded within the standard processor pipeline in the form of specialised execution units. The main feature of a vector unit is the capability of performing a single operation on an aggregated data structure, for instance the arithmetic operations on two arrays of data. The functionalities of the vector units are exposed to the programmer through a dedicated set of instructions. Examples of these additional instruction sets are the Intel SSE and AVX [152], the ARM Neon [258], and the IBM VMX [143]. Data-flow computations can be easily expressed through a data-flow graph that describes data dependencies among the operations. Every time operations are applied on independent data, they can be run in parallel. Modern implementations of such systems are based on multi-core processors augmented with hardware features specifically designed to support the data-flow execution model (e.g., on-chip memory used to store the data-flow graph), and re-configurable cores (generally based on FPGAs), where hardware resources are dynamically configured to exploit fine-grain parallelism [297].

Interconnections and memory sub-system. With the advent of multi-core and many-core processors, designers faced the challenge of interconnecting them in an efficient manner [239]. Interconnections are also key elements in the design of coherence protocols that are used in multi-level cache memory hierarchies. When the number of cores to connect is relatively small, several alternatives are available to designers: from a simple shared bus to rings and crossbar switches. Shared busses cannot scale due to the increased latency emerging from contention and arbitration of the shared medium, the limited number of bits available (bus width), and limits constrained by the electrical properties of ultra-deep sub-micron processes (e.g, wire delay). Rings present a regular architecture, where packets in each node can be injected and ejected by resorting to a simple logic (generally referred to as ring stop) composed of small buffers, a multiplexer, and a demultiplexer. Ring stops are connected one to another forming a closed loop. Thanks to the availability of a large number of wires and pins in modern chips, structured networks wiring (networks-on-chip) have been proposed as a scalable and modular solution for interconnecting larges numbers of cores [121, 67, 300, 68, 32]. Networks-on-chip exploit packet-switched communication mechanisms to transport data, instructions and control messages, similarly to what happens over standard Internet connections. Several topologies have been proposed for NoCs: Fat-Tree (FTree) [247, 291, 40], Torus [367], 2D-Mesh [65], Butterfly [232], Octagon [226].

Cache is crucial for the good performance of the processor. Efficient use of different cache levels helps to achieve a high throughput, but sometimes it becomes a bottleneck for the performance. Cache helps to achieve sharing and dynamic management during the execution. Cache is a smaller set of memories which are faster. Cache memory follows a hierarchical model like L1 (level 1) cache, L2 (level 2) cache, and L3 (level 3) cache. Caches uses the concept of locality of references (temporal locality and spatial locality) to speed up the execution. There are some popular techniques to improve cache efficiency like Prefetching [243], Way Prediction, Data Access Optimization (Loop Interchange/Fusion/Blocking), Data Layout Optimization (Array Padding/Merging/Transpose) etc. In NUMA machines, where each core has its own local caches, inefficient data accesses cause inconsistent data among all the other caches in the whole system. Cache Coherence Protocol is used to maintain all cache copies consistent when contents are modified. The two most popular hardware-based Cache Coherence Protocols are Snoopy Cache Protocol [304, 178] and Directory-based Protocol [355, 96, 41]. Snooping is a broadcast-based protocol but not scalable. Unlike Snooping, the Directory-based protocol is a non-broadcast coherence protocol. Directory-based protocol maintains three cache states (Shared, Uncached, and Exclusive) to store different states (updated, not-updated) of data. It must track all the processors that have shared state data (up-to-date data).

Programming models. Using efficient programming models, we have to harness the inherent parallelism that these systems have to offer. Parallel environment programming is very complex due to the existence of various functional, execution units. Generally, there are three main memory architectures namely shared memory, distributed memory, and hybrid memory models. All current programming models are highly influenced by these memory models. Thread-based programming is based on a shared-memory model. In parallel programming, a single program is a collection of light-weight threads with their own set of instruction segments and datasets. The two most widely used variants of threads are POSIX threads (widely knows as Pthreads) based on the IEEE POSIX 1003.1c standard and OpenMP (or Open Multi-Processing). OpenMP API is a collection of three components: compiler directives, OpenMP runtime library, and environment variables. Programmers can control parallelization during execution, but deadlock occurs when unintended sharing happens. It's based on a fork-join thread execution model. Message Passing Interface (MPI) is a thread-safe, programming model meant for the HPC domain and initially based on a distributed memory model. Distributed shared memory or DSM [282, 320] works by transforming shared-memory accesses into IPC (Inter Process Communication). The Partitioned Global Address Space (PGAS) model [375] views address space globally and effectively exploits locality via efficient abstraction. It inherits the features from both shared as well as distributed memory architectures and are based on a Single Program Multiple Data (SPMD) execution model. In this programming model, the main focus is on dataset operations. There are many program models based on PGAS model; two of them are Unified Parallel C (UPC), which is a PGAS extension of C language [112], and X10[101]. The MapReduce framework, which is highly influenced by functional programming languages like Lisp, is a distributed programming model first introduced by Google [127] for indexing web pages by replacing their original indexing algorithms and heuristics in 2004. This programming model provides a simple yet powerful way to implement distributed applications by abstracting complexities of distributed and parallel programming. Hadoop MapReduce can be implemented by many languages like Java, C++, Python, Perl, Ruby, and C.

OpenCL (Open Computing Language)[351, 279] is an open-source, cross-platform, hardware-centric, heterogeneous, parallel programming language. OpenCL supports both in-order as well as out-of-order execution. It also supports data parallism (primary goal) as well as task-level parallism. In OpenCL terminology, the target platform is called an OpenCL device. The main purpose of OpenCL applications is to efficiently execute OpenCL functions in OpenCL devices. It has two low-level APIs (Platform and Runtime APIs) and kernels to execute program in various platforms (like CPUs, GPUs, DSPs, FPGAs, etc). The Program Execution Model or PXM defines how the application maps onto the hardwares. OpenCL PXM has two parts. One that runs in the host system, and the other runs in devices (called Kernels). Nvidia Compute Unified Device Architecture (CUDA) [290] is a parallel-programming framework to program the heterogeneous model (in which both CPU and GPU are used). Like OpenCL, in CUDA, CPUs are referred as hosts, and the GPUs are referred to as devices. It can access the memory of both hosts and devices. It provides a comprehensive development platform for programmers (C and C++) to accelerate applications by exploiting the GPUs multi-threading capabilities. CUDA has its own compiler, math libraries, and debugging and optimizing tools. In CUDA, hundreds of threads are executed at once, and the scheduler is responsible to schedule the threads.

Moving from programming few heterogeneous devices (shared memory model) to a large number of heterogeneous compute nodes, many efforts have been made, although they are mostly confined to their effective programming. A few of them are MPI+CUDA [211], OpenMP+CUDA [245], and CUDA+OpenMP+MPI [387]. In recent times, the Hybrid MPI+PGAS programming model [219] become very popular for exascale computing.

Hardware-level language or HDL (Hardware Description Language) is very fast, but its code is complex to write. In HDL, programmer codes at circuit/logic gates level, thus he needs very accurate understanding of the system components. In FPGA, domain HDL languages are used to achieve reconfigurable computing. HDL is used mainly to design hardware. Verilog [365] and VHDL (VHSIC—Very High Speed Integrated Circuits—Hardware Description Language) [280] are two flavours of HDL. As already mentioned HDL is very complex for evaluating system-level exploration; we needed tools which can abstract away these complexities. High Level Synthesis (HLS) [166] is a tool which takes an abstract behavioral specification of a system and maps it to the circuit level. Two widely-used commercial HLS tools are Vivado HLS (from Xilinx) and OpenCL SDK (from Intel-Altera).

1.3   Heterogeneous Devices in Modern Data Centers

Modern data centers aggregate a huge number of heterogeneous systems on a large scale. Besides general-purpose processors, many-cores and other specialised devices support the fast execution of large distributed applications. ML and DL applications are also forcing Cloud providers and HPC centers to adopt heterogeneous devices tailored for AI applications. In the following, we summarise the widely adopted technologies that are at the basis of this data center transformation process.

1.3.1   Multi-/Many-Core Systems

Among the variants, the Intel Many Integrated Cores (MIC) architecture received lot of attention from the scientific community as the building block for fast parallel machines. MIC devices offer large performance on applications requiring the execution of floating point operations, along with an easy programming interface. Indeed, MIC is built upon X86 architecture and manufactured using 22-nm lithography; and it shares the instruction-set architecture (ISA) with more conventional X86_64 processors. Xeon Phi is a full system which has full-fledged cores that run Linux. Each core is multithreaded (four threads per core) and follows in-order execution in the previous generations, while introduces out-of-order execution in the last generation. Cores are interconnected by a mesh of bidirectional ring. MIC supports SIMD execution and also specialised SIMD instruction sets such as AESNI, MMX, AVX, or SSE. Apart from that, the accelerator card only understands X87 instructions.

Each core has an L1 cache (of 32 KB for data and instruction), L2 cache (of 512 KB for both instruction and data, totalling 1 MB per tile), but not an L3 cache. All L2 caches have their *tag directories* and TLBs. These distributed tag directories are used to provide uniform access and to track cache-lines in all L2 caches. It supports MESIF protocol for cache coherency and memory coherency. Phi micro-architecture is based on scalar pipeline and vector processing. Each core has two dedicated 512-bit wide vector floating point units (VPUs) and are crucial for the higher performance. It also supports Fused Multiply-Add (FMA) operations, and if FMA is not used, then performance are reduced by half. Each core can execute two instructions (One on U-Pipe and another on V-Pipe) in its core pipeline in every clock cycle. Only one floating point or vector instruction can be executed at any one cycle. The instruction decoder is a two-cycle unit and hence, at least two threads are needed to attain maximum core utilization. In the last generation, the capabilities of each tile (core) have been further increased being able to execute up to 4 threads per core. Experiments on Xeon Phi using the sparse-matrix vector multiplication (SpMV), and sparse-matrix matrix multiplication (SpMM) proved that memory latency is the main concern for the performance bottleneck, not the bandwidth [332].

Other many-core general-purpose architectures explored the VLIW design approach. Two examples of such kinds of architectures are the Kalray's MPPA-256 (a single chip many-core device manufactured using 28-nm lithography [125]) and the Tilera CMP.

1.3.2 Graphics Processing Units (GPUs)

A graphics processing unit (GPU) contains massively parallel small SIMD-based processors whose capability can be exploited by CUDA/OpenCL programming framework. The reason to use GPU is to achieve large-scale parallel-based vector processing. Referring to Nvidia' architecture, each GPU is powered by multiple programmable Streaming Multiprocessors (SM), which are fast and power-efficient. Each SM holds a group of the (more than 100) coprocessors. Each SM has its on-chip memory that can be configured as shared memory and an L1 cache. In addition to the L1 cache, there exists a read-only data cache only during the function execution. Each SM has a dedicated L2 cache memory. Each SM also has a set of registers for high-speed memory. Global memory is used to exchange data between RAM and device via the PCIe bus. Each SM unit features hundreds of single-precision CUDA cores. Each CUDA core has fully pipelined support for single- and double-precision arithmetic (support IEEE 754–2008 standard).

AMD's accelerated processing units (APUs) [74, 76] integrates (in a single SoC) many small, yet powerful GPU cores with less internal memory bandwidth in a chip that works together with 64-bit host CPU without the need of the PCI Express bus which increases data transfer overhead for many GPU-based applications. The integration of GPU cores eliminates the issues of PCIe transfer and small memory size.

1.3.3 Field-Programmable Gate Arrays (FPGAs)

FPGAs are special kinds of chips that are configurable by the end user, but FPGA resources are of a fixed size and have limited flexibility. FPGAs are programmed using high-level synthesis (HLS), circuit schematics or by using a hardware description language like VHDL or Verilog. The most important components of FPGA are configurable logic blocks (CLBs) or logic array block (LAB) and look-up-tables (LUTs). Every CLB consists of multiple LUTs, a configurable switch matrix, selection circuitry (MUX), registers, and flip-flops. LUT is nothing but a hardware implementation of a truth table. LUTs are used to implement any arbitrarily defined Boolean functions in CLBs. LUTs are also used as small memories or small RAM. SRAM blocks are interspersed in the fabric and can be chained together to build deeper wider memories or RAMs. For processing data, FPGA also has hard IPs (such as a multiplier, DSP, Processor). For faster communication, FPGA offers high-speed serial I/Os and I/Os are programmable to operate according to a variety of signaling standards. All state-of-the-art FPGAs incorporate multi-gigabit transceivers (MGTs).

1.3.4 Specialised Architectures

Initially, a DSP was used to process digital signals and only had accessed its local memories. Host core was responsible for coordinating tasks and distributing data. Recently, the features of a DSP have been enhanced significantly for modern multi-/many-core platforms. The SPE (Synergistic Processing Element) cores of a Cell processor is nothing but eight DSPs and can access the main memory via direct memory access (DMA) and also support load and store instructions [199]. The TI Keystone II architecture is a combination of ARM processors (Cortex-A15) with power-efficient floating-point DSPs (TI C66X) [361]. DSPs are well-known for their power efficient execution (high GFLOPS/Watt). Google's Tensor Processing Unit or TPU[1] is a dedicated ASIC chip for a dedicated workload, mainly machine learning such as Tensor-Flow [33]. The TPU is a programmable AI accelerator concentrating on a high volume of low-precision computation. TPU chips support reduced computational precision, which saves a huge amount of transistors for each operation. This reduced precision helps to increase more operations per second at very low power cost. Today, TPUs are already in service to improve Google's Street view.

---

[1] https://cloudplatform.googleblog.com/2016/05/Google-supercharges-machine-learning-tasks-with-custom-chip.html

1.4   Orchestration in Heterogeneous Environments

The management of resources at the infrastructure level and their allocation to the applications represents a major concern to improve energy efficiency, as well as performance and quality of the services provided by modern data centers. As aforesaid, modern data centers leverage on a rich set of hardware and software technologies. Different hardware technologies are at the basis of the large heterogeneity that one can find in modern data centers and provide large benefit in terms of energy efficiency (more specialisation of computing elements for specific tasks) and performance. However, such heterogeneity creates challenges in managing the entire infrastructure, since management software has to understand differences among the various computing elements. The software stack that is responsible for managing infrastructure and allocating resources for the applications is known as an orchestrator.

The OPERA project has developed a software module which provides intelligent allocation of infrastructural resources by (statically) deploying application components (Linux Containers—LCs, and/or VMs) to the most suitable hardware elements, and by (dynamically) scheduling application components migration to achieve energy saving. This solution, namely ECRAE—Efficient Cloud Resources Allocation Engine, is composed of two separate software components, each devoted to static allocation or dynamic optimisation of the workload (see Figure 1.2).

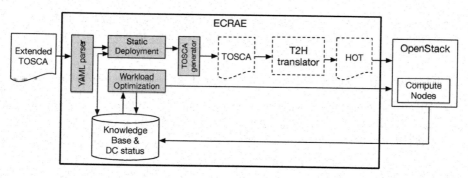

FIGURE 1.2: Logical overview of the Efficient Cloud Resources Allocation Engine (ECRAE): Static deployment and workload optimisation components provide workload management capabilities across the applications life cycle.

ECRAE provides intelligent allocation of infrastructural resources, by (statically) deploying application components (containers and/or VMs) to the most suitable hardware elements, and by (dynamically) scheduling application components migration to achieve energy saving. In addition, a data base referred to as the *Knowledge Base* (KB) provides an updated view of the allocated resources, i.e., for each node it provides the amount of computing, storage, and communication resources already used by running tasks. Specifically, Knowledge Base is based on an internal SQLite relational database and it has two main functions:

- Storing the actual status of the data center controlled by ECRAE orchestrator; at boot time the orchestrator loads the status of the data center to synchronise internal structures with the data stored in KB.

- Storing in a table the matches between tags and flavors. This table is of primary importance for the allocation algorithm.

KB stores the controlled data center resources and status into three tables:

- Host table is dedicated to the host instances, e.g., the representation of the total amount of resources of an OpenStack node. It stores also the power array field, a text string representing the power consumption of the node at different resource utilisation levels. This field is used by the allocation algorithm to calculate the current (energy) power consumption and the projection of (energy) power consumption if a stack is deployed in that node.

- Flavor table stores typical flavor properties and the host where it can be deployed. In OPERA, we assume that a flavor corresponds to only one host, in order to force the deployment of an instance to the specific host found with the allocation algorithm.

- Stack table stores the identifier of a deployed stack and a string that represents a list of flavors belonging to that stack. In this way it is possible to correctly free the allocated resources if the stack is deleted.

In KB there is a fourth table, referred to as Tag:

- Tag table contains all the associations between tags and flavors and is used by the ECRAE allocation algorithm to find the list of suitable flavors matching a tag.

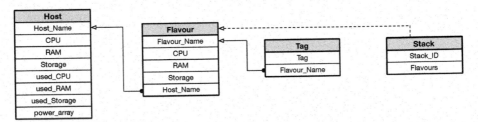

FIGURE 1.3: Knowledge base structure.

In Figure 1.3 are highlighted the dependencies between tables. Flavor table references the host name in table Host. Indeed, a flavor can suit only to one host. Instead, Tag table has the tag-flavor associations. In this table tag and flavor are both primary keys. This means that a flavor can correspond to one or more tags and vice versa. Stack table is detached from others because of the stack object nature. In OpenStack, a stack is an entity that encloses one or more instances and each instance has a flavor. The Stack table stores the identifier and a list of flavor names, in order to know which and how many resources are used by that stack.

1.4.1 Application Descriptor

TOSCA [348] is a standard created by the OASIS committee to describe applications and Cloud infrastructural elements in a portable form. The standard provides a way for creating a template that describes all the required elements needed to correctly deploy the applications. To this end, TOSCA provides a set of standard objects such as node types and relationship types, but it foresees a mechanism allowing object customisation. The possibility to create customised nodes is a key element for the ECRAE orchestration process. Orchestration depends on several fields added to the standard version of TOSCA. With this fields it is possible to better characterise the requirements and the behaviour of the applications/services that will be deployed; specifically, it is possible to choose the most suitable platform according with the defined allocation policies. These customised fields also represent a key feature to enable the integration of containerised applications (using an LXD-based setup) into the OpenStack environment.

1.4.2 Static Energy-Aware Resource Allocation

The ECRAE component responsible for choosing the initial application components placement (e.g., static deployment) is based on a greedy algorithm that aims at finding optimal placing of incoming VMs/LCs also in heterogeneous environments. The main objective is the reduction of the (energy) power consumption in the data center and consequently its energy-efficiency improvement. The (energy) power consumed by a server machine is (mainly) influenced by the (energy) power consumption of the main components such as CPU, memory, storage and network activity. Another important aspect is the power consumption in the idle state, since in such state power consumption can be a large fraction of the peak power consumption [149] [180] [59]. Thus, idle power consumption must be considered in designing the static allocation algorithm. ECRAE allocation algorithm considers this aspect and implements an allocation strategy that tries to allocate resources in the best host from the energy consumption point of view. To this end, the algorithm ranks all the available nodes on their relative efficiency (i.e., power consumption weighted by the current load) and selects the most effective one that can accommodate the requested VM/LC resource requirements (available resources—CPU cores, memory, storage—on servers are made available in the knowledge base integrated in the ECRAE solution). The score is calculated with the following equation:

$$R = (\alpha \cdot C_l + (1 - \alpha) \cdot M_l) \cdot P$$

where the score **R** is the weighted measure of the current power consumption **P** of the node (the power weighted value is biased by the power consumption of the nodes in idle state, so that the P value is given by the power consumption in idle incremented by the fraction due to the machine load), where the weight is expressed by a linear combination of C_l, that represents the CPU load increase expressed as a percentage, and M_l, that represents the memory load expressed as a percentage. The linear combination is obtained by weighting these two load factors with the α parameter, which allow to express how much the application is CPU intensive or memory intensive. In addition, ECRAE considers specific constraints for running the VM/LC; for instance, some VMs/LCs may require the access to a specific accelerator (e.g., FPGA). In that case, ECRAE will (possibly) assign a node that exposes features able to satisfy the such device, reverting to nodes without acceleration[2] only in case no one can accommodate the request.

Algorithm 1 shows the main steps used to select the most efficient nodes where to deploy the application components. Focusing on supporting FPGA acceleration (although, other kinds of accelerators can also be used), first, the procedure evaluates if the VM/LC requires acceleration, e.g., through the use of a dedicated FPGA card (line 2). In that case, the list of hosting nodes equipped with the required device model is used to calculate the rank (line 3). Here, we assume that in case the acceleration board cannot be shared across multiple VMs/LCs, the procedure returns only the list of nodes with the device not in use. To calculate the rank, the cost of using a given node is calculated (lines 19-27). To reduce power consumption, hosts already active are first considered as candidates for allocation (lines 4-5). However, if all the nodes are in the idle state, the procedure will select the less costly node (lines 6-7). Whenever the application component does not require acceleration, the algorithm looks at the list of nodes without acceleration support. As in the previous case, in lines 10-12, the algorithm tries to select a node from the list of active ones. Otherwise, the less costly node will be selected (line 14). To support the selection of the best available node, the same CalculateRanking() cost function is used (see lines 19-27).

1.4.2.1   Modelling Accelerators

To enable the proposed greedy algorithm to take correct decisions when a VM/LC demands for the use of an accelerator, we modelled the performance and power consumption of a generic acceleration device. We measure the speed up in executing a VM/LC on the accelerated platform. Such speed-up value

---

[2]we assume that the application is compiled in such way to run with and without acceleration support

---

Algorithm 1 ECRAE static allocation algorithm

1: **procedure** ECRAEALLOCATIONALGORITHM(hostList, vm)
2:     **if** vm.supportFPGAAcceleration() **then**
3:         $FPGAHostList \leftarrow hostList.getHostSuitableForVm(vm)$ $.getHostWithFPGA()$
4:         **if** $!FPGAHostList.getActive().isEmpty()$ **then**
5:             **return** calculateRanking(FPGAHostList.getActiveHost(), vm)
6:         **else**
7:             **return** calculateRanking(FPGAHostList, vm)
8:         **end if**
9:     **else**
10:         $NoFPGAHostList \leftarrow hostList.getHostSuitableForVm(vm)$ $.getHostWithoutFPGA()$
11:         **if** $!NoFPGAHostList.getActive().isEmpty()$ **then**
12:             **return** calculateRanking(NoFPGAHostList.getActiveHost(), vm)
13:         **else**
14:             **return** calculateRanking(NoFPGAHostList, vm)
15:         **end if**
16:     **end if**
17: **end procedure**
18:
19: **function** CALCULATERANKING(hostList, vm)
20:     **for** host **in** hostList **do**
21:         $actualPowerConsumption \leftarrow host.getPower()$
22:         $CPULoadIncrement \leftarrow vm.getPe/host.getPe$
23:         $score \leftarrow CPULoadIncrement * actualPowerConsumption$
24:         $ranking \leftarrow (score, host)$
25:     **end for**
26:     **return** ranking.minScore().getHost()
27: **end function**

---

is relative and depending on the host CPU with which we are comparing. For instance, we can execute the VM/LC on a X86-64 processor and measure the execution time. Then, we can measure the execution time on the accelerated system (i.e., equipped with the same X86-64); the speed-up is obtained as the ratio between the two execution times. The value obtained by comparing the two execution times is referred to as the *speed-up factor* F_{sp} and can be obtained with the following:

$$F_{sp} = \frac{T_{acc} \cdot W_{cpu}}{T_{cpu} \cdot W_{acc}} = \frac{T_{acc}}{T_{cpu}}$$

where $W_{cpu} = W_{acc}$ is the workload to execute, and T_{acc}, T_{cpu} are respectively the execution time on the accelerated system and the basic one without acceleration. With the previous equation it is possible to get the throughput of an equivalent 'accelerated' CPU, whose execution time is proportional to F_{sp} as follows:

$$T_{acc} = T_{cpu} \cdot F_{sp}$$

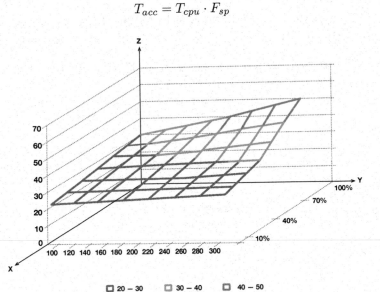

FIGURE 1.4: FPGA energy consumption model.

Unlike the others, providing an abstract but effective power model for the FPGAs is more complex. Indeed, the power consumption of such devices is the summation of two components: i) static power consumption due to current leakage; ii) dynamic power consumption due to the synthesised circuit activity. To correctly model such devices, a 2D dimensional should be kept for sake of understanding parametric model is used, where both clock frequency and percentage of internal allocated resources influence the power drawn. It is worth noting that the toggle activity can be neglected, since it provides a deviation from real power consumption of less than 1%. Figure 1.4 shows an example of such model: the power consumption (z-axis) of a FPGA can be expressed as a function of its clock frequency (y-axis) $[MHz]$ and its internal resources utilization (x-axis) (e.g., logic blocks, look-up tables, DPSs).

1.4.3 Dynamic Energy-Aware Workload Management

Keeping the workload allocation within a data center optimally distributed across the available nodes is mandatory to achieve the best level of energy efficiency and resource usage. Although, static deployment algorithms can provide local optimal allocations, the distribution of applications components, over time, leads the system to a not-optimized state. Not-optimal state may emerge over the time for different reasons. Among those, newly launched applications change the amount of resources consumed (number of required cores, amount of memory, communication bandwidth, storage space), which in turn can lead running applications to experience worse performance and thus to consume more energy. However, by periodically scheduling to launch a dedicated algorithm, a new allocation for the running Linux containers and VMs can be discovered, which is characterized by lower (energy) power consumption.

Addressing this issue is a challenge, since the heterogeneity of hardware and the size of data center infrastructures. To be effective an algorithm must:

- Provide a workload schedule in a short time.

- Be aware of heterogeneity (both in terms of CPU architectures and accelerators).

- Apply to large data center instances (hundreds of nodes).

Additionally, to be beneficial, the used algorithm must try to optimize several parameters, specifically reduce the number of running nodes (consolidation) by also switching off idle nodes, to select the one that provides the largest energy saving, avoid to negatively impact on the application performance and reduce the number of Linux containers and VMs migrations.

Data center resources are huge but not infinite, especially resources on servers can be quickly exhausted. Resources that are available on the servers for running applications can be limited to the CPU cores, main memory, bandwidth and storage. Secondary storage can be easily addressed by assuming the presence of high-end networked storage subsystems. Similarly, with the growing adoption of high-performance interconnections (such as Infini-Band [359, 225] and optical interconnections), we assume that the bandwidth provided by the interconnection is not influencing any workload allocation strategy (we are simplifying the problem formulation by assuming that bandwidth is large enough to accommodate application requests).

The optimal allocation of tasks (e.g., application components, microservices, etc.) on a given group of resources is a well-known NP-hard problem. From the viewpoint of the Cloud Infrastructure Provider (CIP) the allocation problem also involves two aspects: i) the VMs or LCs run for an unknown period of time; ii) the actual profile of the running tasks is unknown. To cope with the first aspect, a snapshot-based approach is used, so that it is possible to remove the time constraint (i.e., a snapshot of the current allocation is periodically extracted and used to feed the optimization procedure).

The problem formulation is based on the optimization model proposed by Mazumdar et al. [271] (i.e., the mathematical formulation can be expressed as a mixed-integer linear programming (MILP) model), and it is intended to allow finding workload schedules with lower power (energy) consumption and with a reduced number of active servers (Server Consolidation Problem - SCP). Hereafter, the workload will be considered as the set of tasks (either VMs or Linux Containers—LCs) that are running on the turned-on servers (data center).

1.4.3.1 Evolving the Optimal Workload Schedule

Evolutionary computation and specifically *evolutionary algorithms* (EAs) are part of the broader domain of artificial intelligence algorithms. At the basis of EAs there is the idea of mimicking the way biological beings evolve towards individuals with a better adaptation to the environmental conditions. The transition to one population instance (i.e., the group of candidate solutions of the problem at a given point in time) to the next is governed by applying *evolutionary operators*, i.e., functions that manipulate the structure of the individuals according to a given rule. EAs are stochastic-based, population-based heuristics, and over time several approaches have been proposed to speed up the search space exploration. One important aspect of such stochastic-based heuristics is that they do not ensure finding the global optimum (i.e., maximum or minimum of the objective function) but are faster than other approaches in exploring the search space. Furthermore, since they are stochastic-based methods, different solutions can be discovered across multiple executions of the algorithm starting from the same initial condition. We will discuss improving stability of the proposed heuristic in the following sections.

At the basis of the *evolution strategy* (ES) algorithm there is still a group of candidate solutions (*individuals*), each representing a possible workload schedule (i.e., a way of assigning tasks to available compute nodes, still fitting in the constraints of the MILP formulation). The exploration of the search space is done by applying, at each iteration of the algorithm, *mutations*. Figure 1.5 shows the main steps composing the evolution strategy algorithm. The basic steps are as follows. Before, starting to explore the search space the population is initialised with the initial configuration, which correspond to the current allocation of the resources (initial workload schedule). Such a schedule is evaluated to calculate the corresponding *fitness* (i.e., in our case it corresponds to the overall (energy) power consumption of the data center (DC)). It follows the main loop, where through several iterations the best candidate solution is extracted. To this end, in each iteration, one of the five mutations is selected according to a certain probability distribution. The

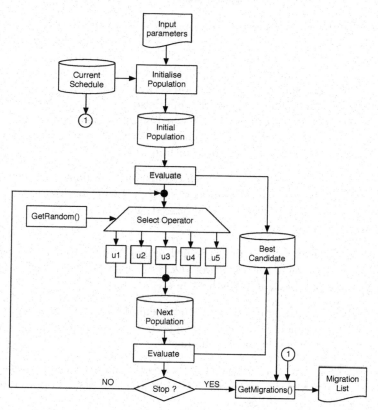

FIGURE 1.5: Flow chart for the evolution strategy algorithm used to find the optimal workload schedule.

operator is applied to each candidate of the population to generate the next ones. Each new candidate is evaluated, and if a new best-ever solution is found, the current solution is updated accordingly. The algorithm stops if one of the *stopping criteria* is met. As stopping criteria, we set the maximum number of iterations of the evolving loop and a given maximum elapsed time. Since the application of mutation operators $u1$, ..., $u5$ is governed by a random process, different runs will use such operators in a different way. Both the thresholds for the stopping criteria, as well as the operator probability distribution are input parameters of the algorithm. Whenever the algorithm exits the main loop, the best-ever found solution (*best candidate*) is compared with the initial allocation (*current schedule*) to determine the list of VMs/LCs to migrate. To ensure a high-quality solutions, the algorithm also applies a final consolidation step: it tries to saturate server resources to increase the number of idle nodes that can be switched off. If the consolidation results in a worse schedule, the best-ever candidate solution is preserved and used. Also, the algorithm tracks the number of generations that have elapsed since the last best-ever candidate found. Whenever that number of generations exceeds a predefined threshold, a steady-state condition is detected. In that case, the current population is discarded, and it is reinitialized (population restart) using the last best-ever candidate solution. Such a new population is then mutated as described above.

Data structure. The proposed ES heuristic uses a dedicated data structure to effectively represent a workload schedule and to keep track of allocated resources in the data center. To represent and efficiently manipulate candidate solutions, each individual is composed of four vectors:

- *Workload* **W**: is an array of length equal to the number of VMs/LCs to allocate. Each element i of the array ($W_i = j$) contains the identification number j of the host where it is allocated, where $j \in [0, N_s - 1]$ and N_s is the number of available hosts.

- *Delta CPU* **Dc**: is an array of length equal to the number of available hosts. Each element i of the array ($Dc_i = c$) contains the number of CPU cores still available (c) on the host i.

- *Delta MEM* **Dm**: is an array of length equal to the number of available hosts. Each element i of the array ($Dm_i = m$) contains the amount of memory still available (m) on the host i.

- *Energy* **E**: is an array of length equal to the number of available hosts. Each element i of the array ($E_i = e$) contains the (energy) power actually consumed (e) by the host i.

ES genetic operators. Bin-packing problem, such as the one concerning resource allocation, requires the implementation of dedicated mutation operators to allow the algorithm to explore the search space. Unlike the basic mutation operation used in ES (in such cases, the ES is intended for continuous optimisation problems), where each element of the solution array is changed by adding a small random quantity (i.e., generally, a random variable with Gaussian distribution is used), here the operators must preserve the correctness of the workload schedule. A workload schedule is valid (correct), if it satisfies all the constraints expressed by the problem model. Given the data structure described above, we designed five different genetic operators that ensure preserving correctness of the candidates:

- Task swapping (TSWP): This operator aims to discover better VMs/LC fitting configurations. To this end, the operator randomly selects one task from the W list and searches the others for a task that it can be exchanged with.

- Task first-fit consolidation (TFFC): This operator aims to find allocation configurations in which tasks are more consolidated on less running servers. To this end, the operator randomly selects one task that can be moved on another server. The server is selected by scanning the list of all servers and selecting the first one that has enough resources to accommodate the moving task.

- Task best-fit consolidation (TBFC): The TBFC operator is very similar to the TFFC operator, in which respect it uses the 'best node' among all the ones that have enough resources to accommodate the moving task. The best node is represented by the one with the largest unused amount of resources, both in terms of CPU cores and free memory.

- Server consolidation (SC): This operator aims to saturate the available resources of a given selected server. The idea behind its use is to make room on other servers for moving in a subsequent iteration larger task. To this end, the operator randomly selects one server on the server list (looking at the **Dc** and **Dm** arrays), and iteratively uses the following condition to move tasks on it:

$$(D_c - R_c) \geq 0 \wedge (D_m - R_m) \geq 0$$

R_c and R_m represent, respectively, the resources (in terms of CPU cores and memory) required by a given task. The tasks to move on the selected server are selected iteratively by scanning the **W** list.

- Server load reduction (SLR): This operator works in an opposite direction compared to the SC operator. In fact, given a randomly selected server i, the operator tries to redistribute the whole load on other servers. To this end, for each task assigned to the selected server, its required resources R_c and R_m are compared to those of other servers (i.e., D_c, D_m for a server $k \neq i$). Thus, the server list is scanned to search for a server k that can accommodate partial or the whole load of server i.

1.5 Simulations

Aiming at evaluating the effectiveness of the static and dynamic allocation algorithms (see Section 1.4.2 and Section 1.4.3), we performed a large set of experiments. Regarding the static allocation algorithm, we used a state-of-the-art simulation framework—*CloudSim Plus* [151] [260] [86], taking into account two different types of workload. Specifically, we described VMs requiring different numbers of MIPS to process (and cores), as well as different amounts of memory. Virtual machines requiring a high number of MIPS to process have been marked for the execution on nodes equipped with an FPGA. The machine configurations in the simulated data center are heterogeneous too. We provided to the simulator machines equipped with processors exposing a variable number of cores and MIPS capacity. To support acceleration, some nodes have been marked as associated with an FPGA board. We considered two type of accelerators (midrange and high-end) to cover a broader spectrum of devices. The heterogeneity of VMs and hosting nodes allowed us to simulate workloads with higher dynamics (more typical for Cloud-oriented applications), and workloads with long-lasting VMs (i.e, HPC-oriented workloads).

To demonstrate the effectiveness of the proposed evolution strategy (ES) heuristic, we implemented the algorithm in a high-level language (specifically, in C for fast execution) and added the capability to generate an initial workload. Then, the algorithm has been tested using different data center configurations, which are aimed at showing scaling capability of the proposed solution. Initial workload generation requires to randomly place virtual machines and/or Linux containers within the simulated infrastructure. During the workload-generation process, different types of VMs/LCs are randomly picked up according to a given probability distribution and assigned to the nodes. The nodes are selected among the ones that have enough resources to host the new VMs/LCs. According to Mazumdar et al. [271], the generation process iteratively selects nodes one at a time and tries to saturate its resources by assigning VMs/LCs. It is worth noting that such a workload generation process is necessary to simulate the data center conditions at a certain point in time, i.e., the current placement of the VMs/LCs on the nodes, which will be optimised.

Various aspects influence the generation process:

- *The distribution of VMs/LCs.* Since there are several types of VMs/LCs, each requiring a different amount of resources, the number of instances may affect the complexity of the allocation process. For instance, a high number of largest VMs/LCs will quickly consume resources on the hosts, leaving less space for the next VMs/LCs to allocate. Conversely, smallest VMs/LCs consume less resources, thus leaving more space for next VMs/LCs to allocate, even in case of larger ones. Since small VMs/LCs are more frequent than larger ones, we set up the probability distribution in such a way instantiation of smaller VMs/LCs is more probable.

- *The resource saturation level.* New VMs/LCs can be allocated to the host whenever there are enough resources to host them. However, in a real data center, hosting nodes are not completely saturated, since it is necessary to reserve some space for management software (e.g., OpenStack compute nodes require hypervisor and networking agent to be installed). To this end, we defined a resource-saturation parameter which defines the fraction of overall resources that can be used to run VMs/LCs. For instance, setting saturation equal to 0.75 means that only 75% of cores and 75% of the memory on the servers are dedicated to support running VMs/LCs. While it is possible to define a separated parameter for different resources (i.e., CPU cores, memory), in our experiment we set them equal to each other.

- *Type of VMs/LCs and type of servers.* Different servers, as well as different types of VMs/LCs to run, greatly influence the way they will be spread over the data center. We defined a set of node types and VMs/LCs types that is representative of real workloads.

- *Optimization model constraints.* The set of constraints expressed in the mathematical formulation of the optimization problem influences the allocation process. The more resources on the nodes that must be constrained (e.g., besides CPU cores and memory, also network bandwidth can be used), the more time is spent in finding a valid node to host a new VM/LC. Therefore we have restricted the actual model to the CPU cores and memory consumption which have been demonstrated [271, 150] to be the most significant factors influencing (energy) power consumption on a node.

1.5.1 Experimental Setup

Table 1.1 and Table 1.2, respectively, show the host configurations used for creating the data center during the simulations, and the types of VMs available to construct the workloads. We described both the hosting nodes and VMs in the CloudSim Plus simulation environment. The hosts are differentiated by means of the number of cores (and MIPS) exposed to the applications (ranging from 2 to 12 cores), and the amount of main memory (ranging from 4 GiB to 12 GiB). For each node type, we provided two subversions: one equipped with a FPGA board, and one without accelerator. While the power model of an accelerator such as GPU or many-core can be derived as a linear function of its loading factor, in case of FPGAs it is more complex. To simplify, we analysed a real application (i.e., convolutional neural networks), and we estimated the power consumption for this design (on the Intel Arria10 midrange device) by fixing the clock frequency to a constant value of 170 MHz (we found that depending on the complexity of the synthesized circuit, the clock frequency spanned from 150 MHz to 350 MHz). This led us to generate the following model (R is the percentage of FPGA resources used by the design):

$$P_{A10} = 37.97 \cdot R + 27.5 \tag{1.1}$$

Equation 1.1 can be used to estimate the power consumption of other application, by fixing the clock frequency to the same value while changing the amount of active resources on the chip. A similar model has been derived for the Intel Stratix10 high-end device, as follows:

$$P_{S10} = 153.4 \cdot R + 110 \tag{1.2}$$

Whenever the FPGA is available, we indicated the speed-up provided (when running the largest supported VM) by the accelerator. Finally, we generated a power model (based on equation 1.1 and equation 1.2) for the nodes, considering the minimum and maximum power consumption levels.

Aiming to evaluate the performance of ECRAE algorithms on different scenarios, we generated two workloads with different numbers of VMs to allocate, while keeping constant the number of hosts (i.e., 500 nodes). The first workload is characterised by a number of VMs that is twice the number of hosting nodes (i.e., 1000 VMs). The second workload aims at saturating the data center capacity, trying to allocate VMs that are ×4 the number of available nodes (i.e., 2000 VMs). Network bandwidth and storage capacity was fixed for all the nodes, as well as for all the VM types. All the simulations have been performed on a Windows-7 (64-bit) machine equipped with 8 GiB of main memory and an Intel Core i7 processor running at 2.5 GHz. CloudSim-Plus requires the Java Runtime Environment (JRE) to correctly work, and in our experiments we used Java version 8.

TABLE 1.1: Host specifications.

Host Config.	MIPS	Cores	RAM	F_{sp}	P_{min}	P_{max}
HP (Xeon 3075)	2660	2	4 GiB	×1	93.7 W	135 W
HP (Xeon 3075)	2660	2	4 GiB	×10	–	–
IBM (Xeon X5670)	2933	12	12 GiB	×1	66 W	247 W
IBM (Xeon X5670)	2933	12	12 GiB	×30	–	–
Intel Arria10	–	–	8 GiB	–	27.5 W	57.9 W
Intel Stratix10	–	–	32 GiB	–	110 W	225 W

TABLE 1.2: VMs specifications.

VM type	Total Instr.	Cores	RAM	Accel.
1	12,500	2	4096 MiB	true
2	2000	1	1740 MiB	false
3	10,000	1	2100 MiB	true
4	500	1	613 MiB	false
5	1500	1	2000 MiB	false

Regarding the dynamic workload optimisation, we assess the capability of the proposed optimisation heuristic by simulating the reallocation (re-balance) of VMs/LCs in a range from 150 to 3250, also varying the size of the data center (ranging from 50 to 1000 nodes). Since the heuristic is driven by a stochastic process, for each run we launched the heuristic multiple times (trials) and we collected all the solutions. At the end of all the trials, we extracted the best overall solution. Each run consisted of 10 trials. During the first trial, we also generated the workload and saved it on disk. The subsequent trials reused the recorded initial workload schedule, since it was used as the initial starting condition. In our experiments, we measured the overall execution time elapsed by the heuristic to complete the exploration of the search space.

To generate the workloads (i.e., the initial workload schedule) we defined 15 types of hosting nodes and 13 types of VMs/LCs. Server nodes had resources in the range of 2 CPU cores and 8 GiB of main memory to 20 CPU cores and 128 GiB of main memory. Power consumption of such nodes were in the range of 30 W to 500 W. Similarly, tasks have been defined in such way that the required resources ranged from 1 CPU core and 1 GiB of main memory to 12 CPU cores and 48 GiB of main memory.

1.5.2 Experimental Results

The ECRAE algorithm performance, i.e., the energy consumption of the data center running different workloads, has been compared with those provided by two largely adopted algorithms, i.e., First Fit (FF) and Best Fit (BF) algorithms. FF aims at providing the fastest decision, quickly allocating an incoming VM to the first node with enough capacity. BF tries to better optimize the allocation by selecting the node having the highest capacity. Both the algorithms do not take into account power consumption of the nodes, thus leading to higher power (energy) usage when compared with ECRAE.

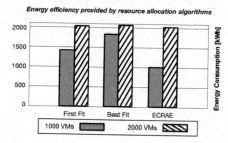

Algorithm	No. Hosts	Incoming VMs	Instantiated VMs	Energy Consumption
FirstFit	500	1000	1000	1425.69 kWh
BestFit	500	1000	1000	1838.87 kWh
ECRAE	500	1000	1000	1006.33 kWh
FirstFit	500	2000	1700	2050.39 kWh
BestFit	500	2000	1700	2071.91 kWh
ECRAE	500	2000	1700	2030.03 kWh

Simulation Results

FIGURE 1.6: Experimental results for the allocation of 1000 VMs and 2000 VMs on 500 hosts, using FF, BF, and ECRAE algorithms.

Static allocation. Figure 1.6 shows the efficiency achieved by the three allocation strategies on different workload conditions. We generated workloads according to the VM types listed in Table 1.2; whenever VMs of type 1 and 3 must be allocated and no one of the machines equipped with an FPGA is available, then a free machine without acceleration support is chosen. This allows FF and BF to correctly search and allocate accelerated VMs on (a subset of) the available machines. From the results, it is clear that ECRAE is always able to take better decisions that minimize the overall energy consumption. On a medium load level (i.e., allocation of 1000 VMs), ECRAE provided 30% of energy saving over the FF solution, and 45.2% over the BF solution. This can be ascribed to the fact that ECRAE does two searches, the first one among running nodes, if there is a running node with enough free resources the algorithms stops. Otherwise it searches among idle nodes. Conversely, BF searches for the less-loaded node for allocation, thus being induced (often) to select idle nodes. Since idle machines may draw down up to 65% of peak power consumption, contribution of BF-selected machines on the overall energy consumption was in most cases higher than in the case of FF and ECRAE. However, FF is not driven by any consolidation-aware mechanism, thus globally selecting a worse mix of nodes if compared to ECRAE solution.

Trying to allocate 2000 VMs caused the data center to reach the saturation point (i.e., actually not more than 1700 VMs were allocated). In this scenarios, all the allocation strategies should show a similar behavior. However, albeit less pronounced, ECRAE results were better than FF and BF solutions. Here, the energy saving was, respectively, 1% and 2%. When compared to the pure performance, ECRAE should be able to access accelerated nodes, thus in most cases providing better performance.

Dynamic workload management. Dealing with the simulation for the dynamic workload consolidation, the maximum execution time has been kept constant in all the experiments to 300s. Through a preliminary set of experiments, we observed that a population (i.e., the set of solutions that are evaluated each algorithm iteration) with eight candidate solutions provided the best trade-off between search-space exploration speed and the overall execution time. In fact, the larger the population is, the more time is required to perform an iteration (i.e., passing from one generation to the next one). Such population size also demonstrated to be effective tackling large problems' instances. For all the experiments we kept the saturation level of the servers (used during the workload generation) equal to 0.720 (here we set the saturation level for CPU cores and memory equal to each other). Such value also represented a further constraint for the ES, which during the solution discovery kept the available resources limited to 72% of the maximum one. The maximum number of generations (representing one of the stopping criteria) has been set up to 100,000. Such value was large enough to explore the search space even in the case of large problem instances (i.e., 2000+ VMs/LCs and 500+ nodes), even though it did not contribute large improvements in the final solution efficiency for small instances. In fact, in small instances, the algorithm converged towards a good solution very quickly. Moving from one generation to the next is accomplished by applying a set of transformations on the candidate solutions, generally referred as a 'genetic operator'. Each operator (kind of transformation) is independent from the others and is applied to a candidate solution following a given probability distribution. Genetic operators receive a fixed likelihood of being used: after an initial tuning such likelihoods have been fixed in a range between 0.12 for the least effective operator and 0.25 for the most effective operator. Figure 1.7 shows the trend of the fitness function measuring the power consumption of the whole data center for the smallest instance (a) and the largest instance (b) of the problem. Besides scalability tests we also tried to compare the behaviour of our algorithm with alternatives, such as the Best Fit (BF) heuristic. However, a comparison with such an alternative approach was not possible since it failed to allocate correctly all the VMs/LCs. In fact, what we note is the general inability of the heuristic to correctly reallocate all the VMs/LCs present in the

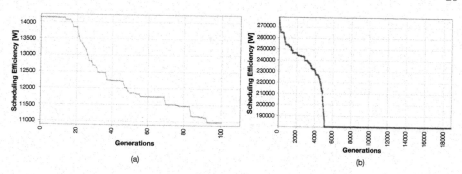

FIGURE 1.7: Fitness function trend (power consumption) for the smallest workload instance (a), and for the largest workload instance (b).

initial workload schedule, due to the absence of a migration mechanism (i.e., a mechanism to reschedule previously allocated VMs/LCs on different nodes). Thus, we consider the initial workload schedule as the reference value against which to compare our (energy) power saving. Table 1.3 shows the results for all the tests performed.

TABLE 1.3: Evolutionary strategy-based workload optimisation results (single thread execution).

Data Center			Power Consumption [KW]		
Nodes	**VMs/LCs**	T_{exec}^{avg} **[s]**	**Initial**	**Optimised**	E_{sav} [%]
50	150	0.81	13.809	3.638	74%
200	637	13.70	55.951	49.062	12%
500	2394	228.54	139.223	70.638	49%
1000	3252	299.98	278.684	180.271	35%

Finally, we studied how the processor architecture running the ES influenced the execution time of the heuristic. We ran the ES algorithm on both the server machine and on an Apple MacBook Pro (2015) laptop. It is worth noting that we executed the algorithms on both the platforms using a single execution thread. Experiments resulted in a larger speedup running the ES algorithm on the server machine. Since the server processor runs at slower speed when compared to the laptop processor, such result can be ascribed to the larger cache memory available on the Intel Xeon class processor. Specifically, the Intel Xeon E5-2630 v4 is equipped with three levels of cache (L1D\$ is large 32 KiB/core, L2\$ is large 256 KiB/core, and L3\$ is large 25 MiB); on the contrary, Intel Core i5-5257U is equipped with three levels of cache memory, but with the last one smaller than in the case of the Xeon processor (i.e., only 3 MiB). Since the ES algorithm data structures can grow quickly as the number of candidate solutions and the problem instance size increase, the larger cache memory in the server allowed to keep more data closer to the processing cores and resulted in an average speed up of 4× when compared to the execution time of the ES algorithm on the laptop.

1.6   Conclusions

This chapter's contribution is twofold: on one hand, it presents an updated view of the heterogeneity in terms of hardware and software support (programming frameworks and models); on the second hand, it highlights the challenges concerning managing such a vast variety of hardware and software at scale, as well as to support applications belonging to Cloud, HPC, and Big-Data domains (i.e., that shows different requirements in terms of computing and memory-capability requirements, as well as different lifetimes of the tasks composing them).

Regarding the latter point the chapter presents a solution for managing the workload in such heterogeneous environments, that has been developed within the EU-funded OPERA research project. This solution, namely the Efficient Cloud Resources Allocation Engine (ECRAE), is responsible for efficiently allocating application tasks on the most suitable DC resources, thus aiming to reduce overall power (energy) consumption. Here, resources are considered as 'computing resources', i.e., the set of server machines available to run such software application tasks (VMs and/or LCs). The ECRAE solution is made of two main elements: *i*) a greedy algorithm to initially allocate resources, depending on the current status of the host servers; *ii*) an evolutionary strategy (ES)-based algorithm for periodic workload optimisation. To this purpose, the servers are ranked depending on their current load and power consumption, and the one that provides the lower (power) energy consumption is selected. Since, servers in a (near-)idle state can consume up to 65%, it requires having the minimum number of active machines to achieve good levels of energy efficiency. To this end, the ECRAE periodically reschedules the application components on the server machines, with the aim of reducing the number of active machines and, at the same time, reducing the overall (power)energy consumption. Finding an optimal schedule for the entire workload is a complex problem (NP-hard), which requires using a smart and effective heuristic to be solved in a short time frame. To this end, we implemented a heuristic based on an ad-hoc evolution strategy, which evolved a group of candidate schedules over time. Experimental results showed the great potentiality in reducing overall energy consumption, by effectively managing heterogeneous infrastructures both in terms of different CPU architectures and the presence of accelerator cards.

Acknowledgments

This work is supported by the OPERA project, which has received funding from the European Union's Horizon 2020 Research and Innovation programme under the grant agreement No. 688386.

2

Modular Operating Systems for Large-Scale, Distributed, and Heterogeneous Environments

L. Schubert, S. Bonfert, and S. Wesner

Ulm University, Ulm, Germany

CONTENTS

Monolithic operating systems: (1) are centralized, (2) need additional effort for distributed computing, (3) cannot deal with the speed of adaptation necessary for modern heterogeneous infrastructures, (4) require more resources than available on embedded devices. Modern Operating Systems need to reflect the application structure better, in the sense that they have to reduce the impact on the environment by focusing only on what is needed locally, rather than trying to cater for all potential cases. Modular Operating Systems offer dedicated "modules" for local execution of distributed and scalable applications with minimal overhead and maximum integration across instances. Inclusion of heterogeneous infrastructures requires only the adaptation of individual modules, rather than the whole operating system. This chapter discusses the architecture, usability and constraints of modular Operating Systems for large-scale, distributed and heterogeneous environments.

2.1 Introduction

Operating Systems (OS) were originally devised to deal with the growing discrepancy between complex resources, fast processing systems and (comparatively) low-demand applications. Instead of executing applications directly on the processor and having to deal with the different I/O systems in code, it offered a mix of libraries to abstract from the resources (drivers) and means to run multiple applications seemingly at once by task-switching (preemption). Over time, processors have become more powerful, devices more heterogeneous and applications more complex, necessitating more complex Operating Systems that can deal with a wide variety of user demands. Implicitly, Operating Systems grew over time and incorporated more and more complexities into a monolithic code block that became more and more difficult to manage as environmental conditions changed—this means that frequently the OS was incompatible with the devices, and driver updates caused the OS to crash etc. Due to the complexity of modern OS, changing just a single functionality can lead to thousands of lines having to be changed [62].

Up until recently, OS developers could rely on constantly increasing processor performance, thus allowing the OS to handle more complex applications whilst becoming more resource-demanding itself. Yet with the introduction of multi-core processors, the environmental conditions changed completely. The increase of performance of a single CPU core stalled, and multiple cores were integrated within a single CPU. Most modern applications adapted to this development by distributing their operation over multiple threads, that are executed in parallel on different cores and thereby removed the limitation to the power of a single core. For the same reasons, Operating Systems were required to become parallel themselves, which introduces the need to synchronize the operation of the individual instances to ensure a consistent system state. On systems with shared memory, this is usually used for synchronization, since it seamlessly integrates into the memory hierarchy. However, monolithic OS frequently use shared data structures for management purposes like, e.g., file system tables or task lists, that hold information about resources in the entire OS, rather than in the individual core. Whenever one of these shared data structures is accessed by different cores, exclusive access and data consistency between caches and memory have to be guaranteed. In most systems entire data structures are locked for exclusive access to solve this challenge. However, this creates a bottleneck for concurrent access to them and may, therefore, depending on the application's behaviour, limit its performance.

For most use cases these approaches work sufficiently well on systems with shared memory. Larger computing systems with distributed memory are generally not considered in monolithic Operating Systems and are treated as completely separate systems, each running their own instance of the OS. The execution of applications on distributed memory systems can be accomplished by explicit communication between its instances that are running on different machines. While this may be assisted by user-space libraries, it still limits the performance of the application and makes its development much more difficult than developing for shared memory systems.

Essentially, modern applications tend more and more towards distributed, heterogeneous setups that thereby benefit less and less from OS support, as they have to start taking distributed resource management, heterogeneous driver setups and data locality into account. At the same time, scheduling and memory management of current Operating Systems do not take such usage into account at all. Instead, they have no awareness of such a distributed, let alone such a heterogeneous environment. For appropriate management of such resources and the applications thereon requires hardware, process and system insights that an application on user space cannot (and should not) have.

It will be well noted that the increasing trend towards virtualization as a means of abstraction from the infrastructure is thereby no feasible alternative as it just hides the underlying complexity rather than allowing its management. As such, a virtualized environment will fail to exploit the system-specific characteristics, let alone coordinate execution of multiple concurrent applications, or, more correctly, to coordinate the execution of all involved distributed processes.

In this chapter we discuss how Operating Systems need to change in order to cater for the needs and circumstances of modern applications and their run-time environments. As noted, this context poses multiple challenges that clash with the traditional approaches to, as well as intentions of, Operating Systems.

We will present an alternative approach to OS design that borrow from modular software architecture concepts. These concepts have been successfully applied to deal with distributed and heterogeneous setups specifically.

This chapter is accordingly structured as follows: section 2 elaborates how modern applications differ from traditional ones and what this implicitly means for their management and why traditional Operating Systems fail to provide for these requirements. Since the knowledge of such problems are by no means new, section 3 will look into other approaches and their drawbacks and benefits, Section 4 will introduce and discuss the structure and organization of modular Operating Systems and how they can contribute to modern applications—with specific focus on catering for distribution, parallelism and heterogeneity. The proposed architecture has been prototypically realized in MyThOS, the results of which we discuss in section 5. MyThOS thereby clearly exhibits higher performance, better adaptability and improved manageability, and thus demonstrates a high potential to cater for the needs of modern applications and infrastructures, as we will briefly discuss in the final section.

2.2 Modern Applications

Traditionally, most applications used to be monolithic blocks that had to be executed on a single (potentially dedicated) machine with little to no internet or network interaction. It is this traditional type of Turing-like applications that defined our whole IT concepts and environments—ranging from processor design to programming models. Obviously, this is far away from our current understanding and development of modern applications: with the introduction and spread of multi-core architectures and the internet, modern applications have to make increasing use of distributed infrastructures, remote data and parallelization-based performance. It is, however, a common misconception that heterogeneity is a new concern, as in fact processor manufacturers have always addressed a wide range of different architectures to address different markets and needs—such as the whole scope of embedded systems and devices. However, traditionally applications have been developed very specifically for these systems and hardly had to be migrated or ported to other contexts—on the other hand, however, Java was developed and conceived for exactly said purposes.

We can note in particular the following aspects as major changes from traditional to modern applications.

2.2.1 Parallelism

Performance is crucial for any modern application—even if just for extended capabilities: the resource requirements constantly increase. Whereas traditionally application developers just had to wait for the next processor generation, which used to be considerably faster, modern processors hardly grow in terms of clock speed anymore. Instead, the number of cores is constantly increasing which requires completely different ways of structuring and developing applications. In combination with the increased distribution, applications have long since left the shared memory region and now also have to cater to explicit communication and synchronization.

Application developers implicitly have to take data dependencies into account (see above) and structure their applications so that the delay for data exchange is taken into account. In the worst case this could even mean that format conversion between different platforms have to be catered for. The problems are well-known for High-Performance Computing developers (see above), who have to develop their applications against hybrid infrastructures constituted of both shared and non-shared memory regions. HPC developers will also be able to tell of the problems of catering for the different processing speeds and implicitly run-time behaviour in heterogeneous setups where hence a single processor can stall synchronization over all application parts.

With standard Operating Systems (and even most modern approaches, see next section), each shared memory system comprises its own full environment—that is to say, it is not aware of the other instances, let alone coordinates its behaviour with them. As a consequence, each respective application part has to be developed as its own logical unit—in other words, as separate processes. Each process is thus locally managed and scheduled concurrently with other local processes and without regard to other remote instances, their dependencies or timing. This can lead to considerable delays, if e.g., a single process has a synchronous dependency.

To manage multiple processes as one application across shared memory systems requires that some data awareness between processes is maintained. The difference between processes and parallel applications is that the latter are constituted of threads that share essential system information—namely specifically memory locations. Using shared memory spaces allows the developer to focus on usage of data rather than having to explicitly cater for data gathering, message exchange and the implicit impact of synchronization (as before). Notably, memory management does not remove the locking/synchronization problem, nor does it mean that no threads would be in a waiting state, but it does relieve the developer from having to cater for explicit message exchange and it allows the Operating System to argue over availability and location of data. Notably, some distributed task management systems, such as Barcelona's Superscalar model, try to identify and exploit this data dependency in non-shared memory distributed systems (see section 2.3).

One specifically important aspect of effective resource management thereby consists in process scheduling, which allows distribution of resources among processes equally, but also, and more importantly, filling waiting time with other sensible processes and tasks, rather than wasting resources. Specifically important for good distributed scheduling is the option that data dependencies can be sensibly coordinated, or that dependent tasks can be scheduled upon data availability. This however necessitates that threads can be instantiated and migrated to other execution cores. Whereas in shared memory systems this is fairly easy and straight forward, distributed memory systems have no contextual awareness of the processes and can thus not easily move processes to a new environment—specifically not, if the system is architecturally different from the original process host.

If a single Operating System were to tackle all these aspects, it would grow considerably in scale and complexity, making it even more unwieldy and hence more difficult to maintain and manage. What is more, though, a single, centralized instance would not be able to scale, let alone deal with the degree of concurrency and distribution.

2.2.2 Distribution and Orchestration

Reuse of existing services and functionalities is increasingly important at a time when the complexity of software surpasses the capabilities of individual developers. What is more, many services are bound to specific providers—and if only for the data associated with it. One important consequence that we have to take home from this is that modern applications consist frequently of distributed services or functionalities which each perform different tasks of the overall application. As a consequence, they also require different execution support and pose different demands towards the resources.

Along the same line, data sources may be distributed all over the internet and the respective data cannot be simply downloaded to a local site for faster access. Invocation of services and data querying requires full knowledge of the interfaces and the means of interaction—to this end, web services and standard communication protocols have been devised. However, these interfaces are not easily exchangeable, requiring constant maintenance and binding the developer to a set of given services—even if the same functionality is offered by an alternative service, too. Similarly, communication and data conversion take a considerable toll on performance.

This leads to similar problems than for parallel applications over non-shared memory systems: data dependencies, scheduling, communication delays all degrade the performance of the application and increase the development effort considerably. Dependencies need to be explicitly encoded and catered for, and there is no way to tell the Operating System which data can be maintained locally in cache, which one needs to be kept consistent at what level etc. All these aspects have to be catered for by the developer instead. This affects equally data sources, remote services and functions: when communication is synchronously bound, in the form of request and response exchanges, execution will stall until the respective data has been received; if, on the other hand, service invocation is bound asynchronously, according event handlers need to be added to the code and according tasks need to be scheduled instead while the service is waiting for response (cf. previous sub-section).

For the Operating System, this is an explicit I/O behaviour, even though it is used in the same fashion as local data access and function invocation.

2.2.3 Dynamicity

A clear consequence of the service-based integration of functionalities consists in the fact that these services are frequently available to multiple users, in some cases even at the same time working on the same data (multi-tenancy). This can lead to data inconsistencies, as well as to access conflicts, respectively to too many requests to handle without overload. Modern systems (here specifically Cloud infrastructures) deal with this by migrating the respective service to a different host, scaling it up or out—which implicitly means that the endpoint is not fixed and can principally lead to unavailability at mid-execution.

For the Operating System, such unavailability would lead to a critical error, considering that data access or code invocation failed. This again is a remainder from traditional process execution, where all data resides locally in the same memory as the code to be executed. From this perspective it makes sense that a failed invocation means that the memory is not maintained correctly and a critical error arose; on the other hand, from service invocations we are used to treating such events as "timeouts" which have to be treated explicitly by the application itself. Obviously, for correct functioning of critical services, the new endpoint has to be identified instead.

In an environment with distributed and potentially dynamic endpoints (see scheduling section, previously), invocations may fail at any time, but should not lead to execution failure—instead, mechanisms for recovery need to be put in place, if such distributed execution is supposed to be possible.

At the same time and vice versa, if an application is supposed to be offered dynamically via the web, the Operating System must support dynamic real-location, as well as concurrent access to data (and functions). Currently, new invocations create additional local processes with a background data access managed by the application, e.g., by using a shared database in the background. As noted, if the system gets saturated with the number of instances, the process will have to be created on another system instead—typically involving explicit serialization of the process state, stopping the process, starting a new process in a new machine (cf. coordination problems, above), to then start the process again and rollback from the serialized state. With traditional processes, such behaviour is not even possible, which gave rise to virtualization in the first instance.

This process becomes even more complicated in heterogeneous environments, where every process will by default be slightly different in terms of data structure and even whole algorithmic organization. By default, also, the new environment will expose different system capabilities via the Operating System, which will have to be incorporated at compile time. Again, abstraction through virtualization (or more correctly, emulation) helps to overcome the compatibility issues; but it comes at the cost of performance and requires according emulators to be in place.

2.2.4   Modern Application's Needs

In summary, modern applications require the following functionalities, which are only partially supported by current Operating Systems:

- Scale agnostic execution (i.e., beyond shared-memory boundaries)

- Hardware agnostic execution (i.e., allowing full, efficient execution, even if the hardware is specialized)

- Location-independent data availability and accessibility

- Location-independent communication/messaging between instances

- Dynamic placement of code and data

In other words, modern applications take a strong "single-system image" view on execution, meaning that they ideally can execute in a distributed fashion completely independent of the resource setup and distribution underneath it.

2.3 Related Work

In traditional monolithic Operating Systems for multi-processor systems, the available resources are managed in a centralized way using data structures shared between the different cores and processors. This architecture is not scalable to infrastructures with several hundreds or thousands of cores, because the overhead for synchronization and coherency protocols dominates the execution time. To reduce this effects, the Corey [73] Operating System allows the application to explicitly control the sharing of data between instances, moving operating system logic to the application and making application development more hardware-specific. A completely decentralized approach is pursued by the multikernel architecture, that was implemented in Barrelfish [63]. In multikernel systems, an independent OS kernel instance is executed on every involved core, regardless of underlying coherency protocols. These instances do not share any data, but interact with each other through asynchronous messages. However, this approach does not offer good adaptability to different hardware or software architectures. The multikernel approach was taken up by several other Operating Systems, e.g., IBM FusedOS [305], mcKernel [172] or Intel mOS [382], that use a hybrid approach of monolithic kernel and multikernel and therefore exhibit similar limitations.

There are several different attempts to create minimal Operating Systems, that only provide a minimal set of functionalities to the application, but are flexible and easily adaptable. One of the first concepts was the microkernel architecture, e.g., Mach [39] or L4 [252], that reduces the overhead of the operating system to a minimum by restricting it to the most basic functionalities like memory and thread management or inter-process communication. Other functionalities like driver or file systems are moved to user-space or left to the application. Due to their small size, microkernels may entirely fit into a processor's cache, which can lead to performance benefits, if Operating System services are frequently used.

This minimalistic approach was carried to extremes by the nanokernel architecture [71]. They delegate additional tasks to user-mode, that traditionally were fundamental parts of the OS like, e.g., interrupt handling, timers, etc., and thereby exhibit an extremely small footprint. Exokernels [144] follow a similar approach by allowing applications to (partially) access hardware directly, bypassing the operating system. While this approach solves the adaptability problem for the kernel, it only moves the issue to another level, since it requires the adaptation of user-space code and does not address the fundamental issue.

Many major Operating Systems like Linux, macOS, FreeBSD, etc. can be extended with loadable kernel modules. They enable the user to dynamically add certain capabilities or device support to the kernel during run-time. Kernel modules can be loaded and unloaded during run-time and may therefore give the impression of a fully modular OS. However, they are only able to extend the base-kernel, not change its behaviour entirely. Furthermore, the base-kernel of these Operating Systems still remains monolithic. The basic challenges that modular operating systems target can therefore not be solved by loadable kernel modules on monolithic OSs.

In academia, modular Operating Systems are an ongoing topic. The factored operating system (fos) [379] dynamically distributes the OS's services across multiple nodes, where a *fleet* of servers offers a specific service. The amount of servers and their location may change during run-time, forming a dynamically distributed OS. Fiasco.OC [240] focuses on embedded systems and offers modularity during compile-time, meaning that the OS can consist of multiple modules that are then compiled into kernel images that are run on the infrastructure. Thus, the OS can be dynamically composed of different modules, but the configuration is fixed during run-time, which enables Fiasco.OC to offer real-time support.

2.4 Modular Operating Systems

As the name suggests, modular Operating Systems are constituted of multiple modules, rather than a monolithic block. Each module is thereby dedicated to a specific OS task and can be realized in different fashions to fulfill different application objectives or meet different infrastructure-specific characteristics. This allows for specialization of the whole execution stack with comparatively little effort, as long as certain key interfaces and functionalities are adhered to. In this section we discuss the key functions and interfaces and derive a general architecture from this.

The primary tasks of an Operating System consist in managing the resources and different applications (respectively, threads thereof) that are to be run on these resources.

In order to perform these tasks, however, the Operating System needs to execute additional functions that are not immediately obvious to the user, such as the process status to allow for preempting threads. Similarly, it needs to monitor the system status not only for accurate timing, but also for the fair distribution of processes over cores—depending on whether we talk about a local or a global view on the system, this involves integration of different information from all processors and a decision across all of these. This means also that the OS must have a representation of the resource it is running on and how it is connected to other instances, so as to make sensible decisions where to execute a new process.

Notably, modern schedulers can reassign cores at run-time of a process, more concretely when it is in waiting state. This is made possible by the cache consistency architecture of modern multi-core processors, which means that a preempted process can start (continue) on another core without impacting too much on the overall performance—though it still requires handling the register file across cores. Obviously, this does not hold true anymore once you leave the shared cache boundary, and the additional cost for serialization and migration needs to be considered before taking a rescheduling decision. We will discuss how a modular Operating System can support and execute such migration, but the actual cost assessment (i.e., the basis for the scheduling decision itself) is outside the scope of this chapter.

Though a strict modular approach would separate all concerns into individual blocks, such an approach is not sensible where the functionalities are co-dependent on each other and separation will only lead to increased effort. It is therefore important that we reassess the functions of an operating systems in light of the capabilities actually needed by a (modern) application. Taking a look at storage management, for example, it is obvious that there is a strong overlap between memory and persistent storage usage—in particular in modern systems, where RAM is swapped to SSD (or even HDD).

2.4.1 Primary Functionalities

From an application point of view, many of the functionalities of an Operating System are not necessary per se. For example, scheduling is only relevant because multiple processes run concurrently—something that an application developer does not explicitly want or should have to cater for. Similarly, memory management and virtual memory primarily arise from such concurrent execution.

As has been discussed in the preceding chapter, modern applications no longer share the same requirements with the ones that gave rise to traditional Operating Systems. Instead, modern applications are distributed, partially shared, mobile, scalable etc. Nonetheless, these applications still rely on the OS to make, e.g., the necessary data available and accessible—this takes a different notion from the traditional memory management for concurrent execution, though.

From the actual expectations of modern applications we can derive the primary functionalities a modern OS—no matter whether modular or not—should support:

Data Availability and Accessibility: Following the concepts of PGAS, the OS should transparently handle the distributed memory and data allocation without the user having to explicitly cater for messages, data consistency or related concerns. Notably, this does not mean that the user/developer should not be able to control data distribution her- or himself, it just means that he should not have to cater for it, if deployment and dynamicity choices are made by the underlying framework. This means first of all that the access to data should be available from all instances that share the data region, independent of the actual location of the data—this is the same concept as was originally introduced with virtual memory and in particular memory swapping, only that it should allow remote access, too.

Data can thus be located (1) within the local cache, (2) the local memory, (3) a local storage device, (4) on a remote instance and there in memory or a storage device. Without additional information, the OS should not address consistency mechanisms or best caching strategies, but needs to allow for according functionalities (see Scale and Distribution Support below). Notably, cache and memory access would not require OS intervention, while the access of remote memory needs support of the OS.

Scale and Distribution Support: Operating Systems and execution frameworks will never make not-scalable code scalable, and it is not their task to do so. Instead, it should ensure that no (or as little as possible) overhead is created by the Operating System in addition to the code's own overhead. It should allow the application to execute across distributed memory systems, without the user knowing the specifics about the underlying architecture. For this purpose, the user has to be able to dynamically create and migrate threads, exchange messages and signals between them and assess the quality of the current deployment during the run-time of the application. Thus, the user can distribute the application's work across the infrastructure, coordinate its execution, detect bottlenecks or spare capacities and adapt the distribution according to that. To conveniently interact with the system, the interfaces should be transparent, so that communication with a local thread behaves identical to communication with a remotely executed one.

Location Independent Invocation: Following what has been said above, not only data needs to be available transparently, but so should functions that are hosted remotely, i.e., on a non-shared memory device. Since functions are scheduled by the respective local operating system, they have no regard of which other instance may or may not need to interact with them. In other words, scale-out may always work against the actual intended performance improvement, if the new instances are further removed from the requesting end-point. No local decision should however take global considerations too much into account, as usage is too volatile to allow for long-term planning. Instead, the system must be able to adapt quickly, thus allowing short-term considerations to drive the scale-out and placement decisions. As a consequence, however, adaptation decisions may break connectivity and should not lead to major performance degradation. This requires that certain information and endpoints are retained, redirected, etc. Notably, as we assume that the service may be invoked multiple times, caching strategies may help to preserve requests and to decide on locality. An important factor to bear in mind thereby is that local adaptability and volatility in a distributed environment can quickly lead to emergent behaviour, where the adaptation of one instance leads to an incremental amount of global adaptations. Stability is therefore a key issue here.

(Some) Hardware Agnostic Execution: Environments will always be heterogeneous and different processors will always lead to different execution behaviour. Heterogeneity exists due to specialization, and hence serves specific needs better than others—rather than enforcing homogeneity, the developer should be enabled to exploit the system specifics without having to cater about the full heterogeneous setup. Instead of exposing a virtual, emulated environment, therefore, the application should be able to interact with the hardware more directly (albeit in a controlled fashion). Current operating systems either hide the infrastructure completely, or offer very hardware-specific system calls to the application—heterogeneity is thus frequently reflected on the OS level. Ideally, however, the operating system just exposes the minimum necessary functions in a hardware-agnostic level and is itself adapted to cater for these independently of the processor specific aspects. These functions should obviously be closely related to the ones identified in this section. All other functionalities should ideally be made available dependent on the system specifics and dependent on need. The latter is an important aspect to which we will come back in the following section.

It will be noted that the essential functionality is still very similar to the ones offered by traditional operating systems, but that they have to support these functionalities in slightly different fashion, i.e., with an eye on distribution and adaptation. Nonetheless, "traditional" requests must still be met, so that even though, e.g., I/O and graphical support are subsumed under the other functionalities, they still need to be provided for explicit application requests.

2.4.2 Architecture

A modular Operating System exposes itself to the application (and the application developer) generally in much the same way as a "regular" operating system would. The main difference, as we shall discuss further below, consists in how the parts of the operating system are aligned and interact with each other to realize the requested behaviour. We can thus distinguish between four main layers with respect to application execution (cf. Figure 2.1):

Synchronous Layer: This is the lowest, most basic layer of the architecture. Ordinary C++ objects are used to implement fundamental data structures and basic hardware abstraction components. These objects are used via the ordinary C++ method and function calls, and their implementation is executed immediately in the current logical control flow.

This layer is responsible for providing the run-time environment that is needed by the asynchronous layer. This includes the scheduling of asynchronous activities across hardware threads and mechanisms for concurrency and locality control. Tasklet objects provide the abstraction for asynchronous

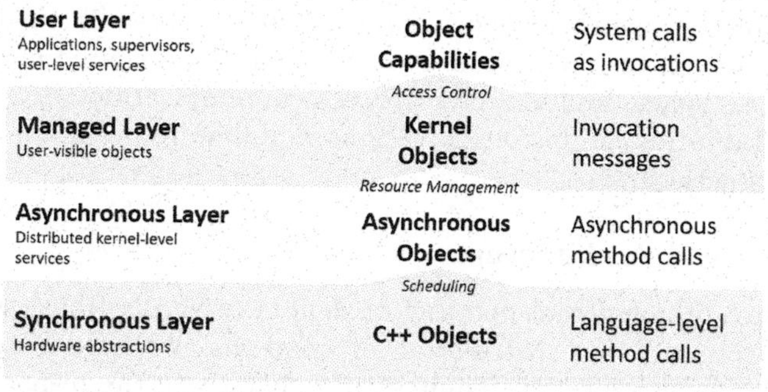

User Layer Applications, supervisors, user-level services	Object Capabilities	System calls as invocations
	Access Control	
Managed Layer User-visible objects	Kernel Objects	Invocation messages
	Resource Management	
Asynchronous Layer Distributed kernel-level services	Asynchronous Objects	Asynchronous method calls
	Scheduling	
Synchronous Layer Hardware abstractions	C++ Objects	Language-level method calls

FIGURE 2.1: Logical layers of a modular OS environment.

activities and monitor objects implement asynchronous synchronization policies such as mutual exclusion—they are conceptually identical to threads but contain their own execution context information.

Asynchronous Layer: On top of the run-time environment from the synchronous layer, this layer hosts asynchronous objects that provide communication via asynchronous method calls. In contrast to arbitrary C++ object methods, the asynchronous methods consume a Tasklet as small-state buffer and a reference to an asynchronous response handler object. The actual implementation of the called method can be executed at a later time and also on a different hardware thread. The call returns by calling a respective asynchronous response method, passing along the Tasklet.

This layer implements shared kernel infrastructure and is not directly visible to the user. Examples are the resource inheritance tree and supporting asynchronous objects for the implementation of kernel objects.

Managed Layer: This layer manages the resources and life cycle of kernel objects through capabilities as an unified smart pointer mechanism. The capabilities track the resource inheritance beginning from memory ranges over kernel objects allocated in this memory to derived weak references that point to the same kernel object. In cooperation with the memory management this inheritance enables the clean deallocation of kernel objects and recycling of the respective memory. Alongside the basic state information that is used by the kernel's resource management, the capabilities include generic and type-specific access rights. These are used by the system call interface to restrict the user's capability invocations. Instead, such invocations are treated like asynchronous messages (or method calls) that are forwarded to the invoke() method of the targeted kernel object.

This layer contains all user-visible and call-able system services (essential functions, see below). A small set of system calls allows push capability invocations from the application to the kernel objects.

Application Layer: Applications, supervisors and other system services live in the application layer. Application threads interact with kernel objects via capability pointers, which are logical indexes into the thread's capability space. Communication on the application layer is possible via shared memory in the logical address spaces and via inter-process communication (IPC) services of the kernel. Libraries can introduce various middleware layers for higher-level parallel abstractions.

2.4.2.1 Logical Architecture

We have already noted that one way of dealing with distribution consists in modularization and thus distribution of concerns. With such an approach, adaptation to new conditions requires only changes in the respective module rather than in the whole monolithic block. An important side effect of this approach is, however also, that distribution of modules requires some form of connectivity between them and hence that their location needs to be tracked and maintained. Another major advantage of modular approaches consists in the replacability and extensibility of the modules themselves, which allows the developer to define new modules and add them to the OS as the need arises, as well as to provide different "flavours" thereof to accommodate for different infrastructures and environments. However, in order for all modules to be available and usable, the interface needs to be well specified. In other words, by using a modular approach we run into the same problem of interoperability as mentioned in the context of service invocation above. As we shall see, however, we think that this drawback is well worth the advantages.

Essential functions are functions that should be provided by all operating systems at an utmost minimum. These functionalities are generally needed to execute applications concurrently and in a distributed fashion, as well as to realise the modularity in the first instance. As discussed in more detail below it should be noted that the modular approach means that not all functionalities need to be present in the same fashion in all OS instances—it is this what makes the OS scalable and adaptable.

The essential functions are:

- memory management: creation, destruction, read and write. Maintains an application-wide instances table.

- consistency management: replication, synchronization. Can be automatically enacted on different memory regions.

- communication: send, receive, asynchronously and synchronously. Needs to include event handlers.

- storage: read and write large amount of data to persistent storage.

- other IO: allows interaction with the user—we will not elaborate this in detail.

- process management: instantiation, preemption, migration, destruction, scheduling strategy.

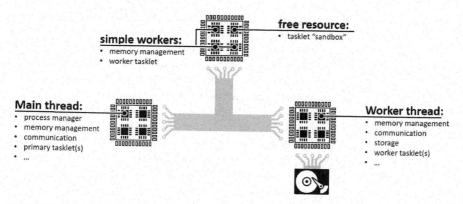

FIGURE 2.2: Exemplary distributed deployment of the operating system modules.

As opposed to traditional Operating Systems, the functions are exposed as individual modules, which thus have to be explicitly invoked in a messaging-like fashion, rather than having to be invoked in memory. The reason for this is two-fold: on the one hand, this allows the easy replacement of modules without having to re-compile the whole code; on the other hand, and more importantly, this allows for free distribution of modules across the infrastructure without having to co-locate all modules with each other.

2.4.2.2 Deployment Architecture

The main point of the modular structure is that no instance in the environment (i.e., no processor and hence no thread) needs to host the full operating system instance, as is the case for monolithic operating systems. Just like a distributed application, each service (or here: functionality) can be hosted on a different machine and will be invoked through the communication channel in the operating system (cf. Figure 2.2).

Each application can thereby either fully control which modules need to be co-located with which application threads (Tasklets), for example by using a graph-like dependency description. Principally, however, the instance control can be left to the operating system itself, using cache-like access logic to migrate the modules on demand: as long as the migration control and the communication modules are present, any module can be handled like a Tasklet itself and thus the operating system's layout can be dynamically reorganized at run-time.

In order to enable such communication, the operating system utilizes hardware-based system call as supported by the processor. This allows initial communication channels between applications and system services and between applications. Direct communication between applications can then be established by mapping shared memory or memory-mapped communication devices into the application's logical address space.

Conventional shared-memory kernels implement this communication through state-changes that eventually become visible to other threads by observing the shared memory. This style requires the use of mutual exclusive execution sections or lock- and wait-free algorithms. Instead we use dedicated message channels between the hardware threads. Messaging-based communication and wait-free algorithms can increase the asynchronicity because it is not necessary to wait by blocking for the completion of emitted tasks. This leaves more processing time for actual computations.

Any communication attempt is intercepted to be matched against the component deployment table at boot-up time. Messages are forwarded to the last known endpoint. Should the endpoint not have the respective instance locally anymore, the invoked instance will return the endpoint to which the respective module was migrated to. It may also forward the request accordingly. This concept builds up on principles from DNS. This ensures that updates are automatically distributed to all interacting endpoints without having to broadcast all changes every time.

We have shown (below) that these mechanisms allow for much faster thread maintenance, which reduces thread instantiation roughly by factor 5. With the fixed interface description on the OS module level, it becomes furthermore possible to interact with different Tasklets irrespective of their concrete implementation—in other words, the OS can host a setup of heterogeneous Tasklets dedicated to specific processors in a mixed overall environment and still allow full communication and interaction with them. Migration can thus switch between different instantiations without having to cater for a completely different execution context.

2.5   Analysis

To evaluate some key performance measures of a modular operating systems, we have developed a prototypical implementation, called MyThOS. The development of MyThOS was partially funded by the Federal Ministry of Education and Research under contract number 01IH13003. This OS is based on a microkernel architecture and features a reduced interface. It is intended to be used in high-performance computing and embedded systems with different hardware architectures. The individual instances of MyThOS can be configured differently from a set of modules to utilize available hardware features and allow the user to adapt its operation to specific application requirements.

Whenever a userspace application intends to use services offered by the operating systems, it issues one or more system calls. In multicore systems this may happen fairly often, because of the dynamic sharing of resources between different CPU cores. In MyThOS, the endpoints of system calls are objects residing in kernel space, which are called kernel objects. These may have a fixed location assigned to them, where each corresponding system call is routed to or migrates to the location where a system call is issued. The former strategy offers high temporal locality because data required within the kernel object may stay in cache at a dedicated location, while the latter strategy offers better load balance. A mobile kernel object is moved to the location of the system call whenever it is currently not executed somewhere else. Otherwise it continues its execution at the current location to improve code and data locality. Services that one operating system instance offers to other instances (e.g., I/O) are implemented in the same way and offer similar flexibility.

We conduct benchmarks on the system-call performance of MyThOS for both fixed-location and mobile kernel objects. For this purpose a dummy kernel object that performs a configurable amount of work is used. From the user space, this object can be called as reported in listing 1.

Listing 1—MyThOS object call in user space

```
1 int workload = 10000;
2 obj->work(workload);
```

After crossing the system call boundary and routing the call to the respective object in the kernel, the call ends up in the respective function in within the dummy kernel object in kernel space as shown in listing 2.

Listing 2—MyThOS object management in kernel space

```
1 void Dummy::work ( int workload )
2 {
3     for ( int i = 0; i < workload ; i ++);
4 }
```

Figure 2.3 shows the measured latency from the invocation of a system call until control is handed back to the application. Note, that this does not imply that the system call was already processed, since processing occurs asynchronously, but the application can continue with its execution. For calls to a kernel object that is located on the same hardware thread as the caller, we measure an average response time of 2500 cycles. This represents the costs of a simple system call. In the case of remote system-call execution, additionally endpoint resolution and communication costs play a major role, which is why the response times are much higher in these cases. For system-call execution on a different hardware thread on the same core we measure a latency of 102,000 cycles, which is only a little lower than in the case of execution on a different core with 106,000 cycles. The ability for the user to specify the location of execution for a kernel object is a major difference between MyThOS and Linux, since Linux does neither support such fine-grained control over the

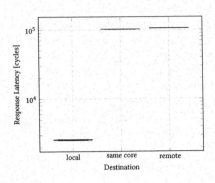

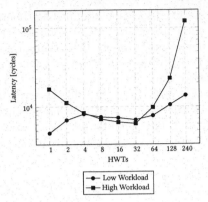

FIGURE 2.3: System-call latency for different types of endpoints (fixed-location kernel object).

FIGURE 2.4: System-call latency for different types of endpoints (mobile kernel object).

execution of system calls nor the remote execution of them. Therefore, no fair comparison to Linux can be made in this respect.

Figure 2.4 shows the response latency for system calls to mobile kernel objects in relation to the number of threads (running on different hardware threads) calling them concurrently. If only a low workload is executed within the system call, the corresponding object is nearly always executed at the location where the system call was issued. Therefore, response times are relatively low. When executing higher workloads inside the system call, the probability of migration decreases with increasing concurrency, and it ultimately behaves like a fixed-location kernel object.

This implementation of system calls clearly demonstrates the advantage of being able to implement functionalities in different manners for different applications. One application may benefit from locally executed system calls, especially if they are only short-running. However, other applications may achieve higher performance with remotely executed long-running system calls and their asynchronous execution.

Due to the slim design of MyThOS and its design dedicated to high scalability, the overhead of common operations is reduced to a minimum. One example of this is the creation of new threads within the OS. We therefore measured the duration of thread creation in MyThOS. The results are shown in Figure 2.5. On average, the creation of a new thread in MyThOS including all supporting structures takes around 6000 CPU cycles. In this case, all necessary operations can be executed on the local core. In contrast, for deployment of the thread to a remote hardware thread, additional communication is required. The deployment time also includes the time for the transmission of a kernel-internal token to the destination and back to the initiator to return the result of the operation. Therefore, the deployment is significantly more costly and takes ca. 12,000 cycles. On the destination hardware thread, code

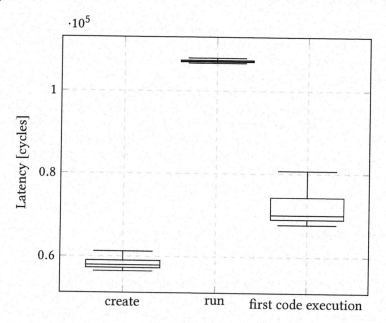

FIGURE 2.5: Thread creation latency.

can be executed right after the token arrived and before its transmission back to the sender. For the time elapsed until the first code is executed in the newly created thread, we measured 7,000 cycles. For comparison, we measured the duration for the creation of a new thread until the first code execution in Linux using pthreads. Similar to the MyThOS case, this includes the setup of all required data structures as well as the deployment of the thread. On average, this operation takes 36,000 cycles in Linux on the same hardware platform. This gives MyThOS a performance gain of factor 5.

While these measurements demonstrate some of the possible performance benefits when using a modular operating system such as MyThOS, it also imposes some drawbacks on developers and users. The possibility to adapt the configuration to specific hardware features, architectures or application requirements yields the opportunity to achieve significant performance gains without changing the underlying infrastructure. To actually achieve these benefits, the developer of an application has to have a much deeper understanding of the internals of the application, its computational behaviour, communication patterns and memory usage. Furthermore, deep knowledge of the available hardware and its features, the system's interconnect and memory hierarchy are required to be able to map the application's operation to the underlying hardware as efficiently as possible. While modular operating systems can assist the developer with the tools to achieve good mapping, the fundamental decisions still have to be taken by the developer. However, these decisions can often be done iteratively by optimizing different aspects one after the other. Thereby, performance gains can be achieved with reasonable effort.

Most modular operating systems do not currently offer fully Linux (or POSIX) compatibility, since this support would require compromise on scalability or flexibility. Instead, they feature different interfaces for thread and memory management, I/O, signals or communication. The porting of an application from Linux to such an operating system (or even between two of them) requires considerable effort. Not only the interfaces may be different in their syntax, but the operating systems may utilize entirely different concepts and paradigms, like asynchronous system calls.

2.6   Conclusion

In this chapter we showed, why modern applications require new Operating System architectures, how modern OSs are limited in their performance and adaptability and gave an overview of the architecture of a modular Operating System and its performance.

Monolithic OSs have been around for many decades and were able to efficiently utilize computers. However, both the hardware and applications changed significantly in recent years. The hardware is increasingly heterogeneous, while applications require more resources and are executed distributedly on many nodes in parallel. Our performance evaluation shows that modular Operating Systems are able to tackle these challenges and offer a significant performance benefit over monolithic ones. Furthermore, they are able to support heterogeneous execution platforms by providing specialized implementations for modules and dynamically adapt to changing hardware and software requirements. Since they are easier to maintain and adapt, we consider modular Operating Systems a promising approach for the future.

3

Programming and Architecture Models

J. Fumero, C. Kotselidis, F. Zakkak, and M. Papadimitriou
The University of Manchester, Manchester, UK

O. Akrivopoulos, C. Tselios, and N. Kanakis
Spark Works ITC, Derbyshire, UK

K. Doka, I. Konstantinou, I. Mytilinis, and C. Bitsakos
National Technical University of Athens, Athens, Greece

CONTENTS

Heterogeneous hardware is present everywhere and in every scale, ranging from mobile devices, to servers, such as tablets, laptops, and desktop personal computers. Heterogeneous computing systems usually contain one or more CPUs, each one with a set of computing cores, and a GPU. Additionally, data centers are currently integrating more and more heterogeneous hardware, such as Field-Programmable Gate Arrays (FPGAs) in their servers, enabling their clients to accelerate their applications via programmable hardware. This chapter presents the most popular state-of-the-art programming and architecture models to develop applications that take advantage of all heterogeneous resources available within a computer system.[1]

3.1   Introduction

Contemporary computing systems comprise processors of different types and specialized integrated chips. For example, currently, mobile devices contain a Central Processing Unit (CPU) with a set of computing cores, and a Graphics Processing Unit (GPU). Recently, data centers started adding FPGAs into the rack-level and switch-levels to accelerate computation and data transfers while reducing energy consumption.

Figure 3.1 shows an abstract representation of a heterogeneous shared-memory computing system. Heterogeneous shared-memory computing systems are equipped by at least one CPU, typically—at the time of writing—comprising 4 to 8 cores. Each core has its own local cache, and they share a last level of cache (LLC). In addition to the CPU, they are also equipped with a set of other accelerators, normally connected through the PCIe bus. Such accelerators may be GPUs, FPGAs, or other specialized hardware components.

---

[1]Part of the material presented in this chapter is included in Juan Fumero's PhD thesis [162] with the permission of the author.

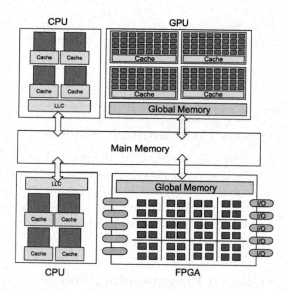

FIGURE 3.1: Abstract representation of a heterogeneous computing system. (© Juan Fumero 2017.)

On one hand, GPUs contain hundreds, or even thousands, of computing cores. These cores are considered "light-cores", since their capabilities are limited compared to those of CPU cores. GPUs organize cores in groups. Each group contains its own cache memory. These groups are called Compute Units (CUs) or Stream Multiprocessors (SM).

FPGAs, on the other hand, are programmable hardware composed of many programmable units and logic blocks. In contrast to CPUs and GPUs, FPGAs do not contain traditional computing cores. Instead, they can be seen as a big programmable core in which logic blocks are wired at runtime to obtain the desired functionality.

Each of these devices, including the CPU, has its own instruction set, memory model and programming models. This combination of different instruction sets, memory models, and programming models renders a system heterogeneous. Ideally, programmers want to make use of as much resources as possible, primarily to increase performance and save energy.

This chapter describes the prevalent programming models and languages used today for programming heterogeneous systems. The rest of this chapter is structured as follows: Section 3.2 discusses the most popular heterogeneous programming models that are currently being used in industry and academia; Section 3.3 shows programming languages that are currently used to develop applications on heterogeneous platforms; Section 3.4 analyses state-of-the-art projects for selecting the most suitable device to run a task within an heterogeneous architecture environment; Section 3.5 describes new and emerging heterogeneous platforms that are currently under research; Section 3.6 shortly presents ongoing European projects that perform research on heterogeneous programming models and architectures; finally, Section 3.7 summarizes the chapter.

3.2   Heterogeneous Programming Models

This section gives a high-level overview of the most popular and relevant programming models for heterogeneous computing. It first describes the most notable directive-based programming models such as OpenACC and OpenMP. Then, it describes explicit parallel and heterogeneous programming models such as OpenCL and CUDA. Furthermore, it discusses the benefits and the trade-offs of each model.

3.2.1   Directive-Based Programming Models

Directive-based programming is based on the use of annotations in the source code, also known as `pragmas`. Directives are used to annotate existing sequential source code without the need to alter the original code. This approach enables users to run the program sequentially by simply ignoring the directives, or to use a compatible compiler that is capable of interpreting the directives and possibly create a parallel version of the original sequential code. These new directives essentially inform the compiler about the location of potentially parallel regions of code, how variables are accessed in different parts of the code and how synchronization points should be performed. Directive-based programming models are implemented as a combination of a compiler—often source-to-source [55, 394, 395, 251]—and a runtime system. The compiler is responsible to process the directives and generate appropriate code to either execute code segments in parallel, transfer data, or perform some synchronization. The generated code mainly consists of method invocations to some runtime systems that accompanies the resulting application. These runtime systems are responsible for managing the available resources, transferring the data across different memories, and scheduling the execution of the code [303, 370, 360, 187, 77, 78].

3.2.1.1   OpenACC

Open Multi-Processing (OpenMP) [296] is the dominant standard for programming multi-core systems using the directive-based approach. Following the successful case of OpenMP, the OpenACC [293] (Open Accelerators) standard was proposed at the Super-Computing conference in 2011. This new standard allows programmers to implement applications for running on heterogeneous architectures, such as multi-core CPUs and GPUs, using the directive-based approach.

To use OpenACC, developers annotate their code using the `#pragma` directives in C code. Then, a pre-processor (a computer program that performs some evaluation, substitution, and code replacements before the actual program compilation) translates the user's directives to runtime calls that identify a code section as a parallel region, and prepares the environment for running on a heterogeneous system.

Listing 1 shows a simple example of a vector addition written in C and annotated with OpenACC directives. Computing kernels are annotated with the `acc kernels` directive. In OpenACC, developers also need to specify which arrays and variables need to be copied to the parallel region (i.e., are read by the kernel) and which need to be copied out (i.e., are written by the kernel). In this example, the arrays `a` and `b` are marked as `copyin` since they are read by the annotated for loop. Since the statement within the `for` loop writes the result of each iteration in the array `c`, the latter is marked as `copyout`. This information is used by the compiler and the runtime system to generate and execute only the necessary copies, to and from the target device (e.g., a GPU). Note that OpenACC requires setting the range of the array that a kernel is accessing. This essentially enables different kernels to access different parts of the same array with the need of synchronization as long as the two parts do not overlap.

Listing 1—Sketch of code that shows a vector addition in OpenACC

```
1  #pragma acc kernels copyin(a[0:n],b[0:n]), copyout(c[0:n])
2  for(i=0; i<n; i++) {
3      c[i] = a[i] + b[i];
4  }
```

3.2.1.2 OpenMP 4.0

Following the success of OpenACC for heterogeneous programming on GPUs, OpenMP also included support for heterogeneous devices since the standard OpenMP 4.0, one of the latest versions of the standard at the time of writing this book [296]. This version of the standard includes a set of new directives to program GPUs as well as multi-core CPUs.

Listing 2 shows an example of how to perform vector addition to target GPUs with OpenMP 4.0. The first observation is that OpenMP for GPU programming is much more verbose than OpenACC. This is because OpenMP is completely explicit about how to use the GPU from the program directive level. OpenMP 4.0 also introduced new terminology for extending the existing OpenMP for multi-cores and shared computing memory systems to be able to express computation for GPUs.

Breaking down Listing 2, line 1 sets a new OpenMP region that targets an accelerator through the `target` directive. An accelerator could be any device, such as a CPU or a GPU. Then, it specifies how the data should be seen by the host (main OpenMP thread on the CPU). By default, data is owned by the target device. Therefore, to copy the data from the host to the device, OpenMP uses the directive map with the parameter to. On the contrary, to perform a copy out, we use the map-from clause. These two clauses specify how to move the data from the host to the device and back.

The second pragma in Listing 2 expresses how to perform computation on the target device (line 3). The target directive indicates that execution needs to be relocated (e.g., moved to a GPU). The teams directive tells the compiler that parallel code should be handled by at least one team (group of thread working together on the target device). The distribute directive tells the compiler that the i induction variable in the loop should be distributed across the teams. The parallel directive activates all the threads inside a team. These threads should run in parallel. The for clause tells the compiler that the work is shared across all the teams. Finally, the schedule(static,1) tells the scheduler that blocks of threads use contiguous memory, to increase memory coalescing.

As seen, OpenMP can get very verbose, giving fine control of the execution to the programmer. Furthermore, programmers have control of how data are shared, allocated, and transferred, as well as how execution is performed and scheduled on the GPU.

Listing 2—Sketch of code that shows a vector addition in OpenMP 4.5

```
1  #pragma omp target data map(to:a[:n],b[:n]) map(from:c[:n])
2  {
3    #pragma omp target teams distribute parallel for \
4    schedule(static, 1)
5    for(i=0; i<n; i++) {
6      c[i] = a[i] + b[i];
7    }
8  }
```

Considerations:

Both OpenMP and OpenACC work for C/C++ and Fortran programs, and they require parallel programming expertise as well as good understanding about the target architecture to achieve performance. To build heterogeneous applications with OpenMP and OpenACC, programmers need a good understanding of the data flow in their application. They also need to understand the different barriers (explicit and implicit), how to perform synchronization, how to share variables, how data are moved and when the data are moved from CPU to GPU and vice versa, and how execution is mapped to the target architecture. Furthermore, these two models, although targeting heterogeneous architectures, and more precisely GPUs, are not equivalent. OpenMP extended the standard that targets multi-core CPUs, meanwhile OpenACC was designed from scratch having GPU computation in mind.

Due to the simpler directives of OpenACC, some programmers rewrite OpenMP programs that target multi-core CPUs using the OpenACC directives to port their applications to heterogeneous devices with the least possible effort. However, considering the overheads of the data transfers to and from the target device, along with the limitations that GPU cores impose, blindly porting OpenMP annotated applications targeting multi-core CPUs to OpenACC may not yield performance improvements. On the contrary, it might even end up giving a worse performance than the sequential execution.

3.2.2 Explicit Heterogeneous Programming Models

This section describes the most common explicit heterogeneous programming models available at the moment, both from the industry and academia. These programming models are relatively new compared to the directive-based ones, such as OpenMP.

OpenCL (Open Computing Language) [294] is a standard for heterogeneous computing released in 2008 by the Khronos Group. The OpenCL standard is composed of four main parts:

Platform model: it defines the host and devices model in which OpenCL programs are organized.

Execution model: it defines how programs are executed on the target device and how the host coordinates the execution on the target device. It also defines OpenCL kernels as functions to be executed on the devices and how they are synchronized.

Programming model: it defines the data and tasks programming models as well as how synchronization is performed between the accelerator and the host.

Memory model: it defines the OpenCL memory hierarchy as well as the OpenCL memory objects.

The rest of the section explains each of these parts in more detail.

Platform Model

Figure 3.2 shows a representation of the OpenCL platform model. OpenCL defines a host (e.g., the main CPU) that is connected to a set of devices within a platform. A host can have multiple devices. Each device (e.g., a GPU) contains a set of compute units (group of execution units). In its turn, each compute unit contains a set of processing elements, that are, in fact, the computing elements in which code will be executed.

The number of computing units (CU) and processing elements (PE) within a device depend on the target device. For instance, if the target device is a CPU, a CPU contains multiple cores. Therefore each core is a CU. In its turn,

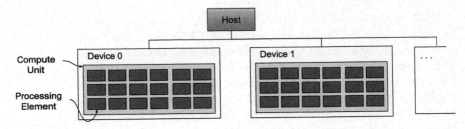

FIGURE 3.2: Representation of an OpenCL Platform. (© Juan Fumero 2017.)

each core contains a set of SIMD (Single Instruction Multiple Data) units. Those units are the PE within a CU.

In the case of GPUs, each CU comprises a set of stream multiprocessors (using the NVIDIA GPU terminology) that contains a set of physical cores, instruction schedulers and a set of functional units. Each small core is mapped to a PE, where instructions will be executed.

Execution Model

The execution model is composed of two parts: the host program and the kernel functions.

Host Program: It defines the sequence of instructions that orchestrate the parallel execution on the target device. The OpenCL standard provides a set of runtime utilities that facilitate the coordination between the host and the device. A typical OpenCL program is composed of the following operations:

1. Query the OpenCL platforms available.

2. Query all devices within an OpenCL platform that are available.

3. Select a platform and a target device.

4. Create an OpenCL context (an abstract entity in which OpenCL can create commands associated with a device and send those commands to the device).

5. Create a command queue to send requests, such as read and write buffers and launch a kernel on the device.

6. Allocate buffers on the target device.

7. Copy the data from the main host to the target device.

8. Set the arguments to the kernel.

9. Launch the a kernel on the device.

10. Obtain the result back through a copy back (read operation) from the device to the host.

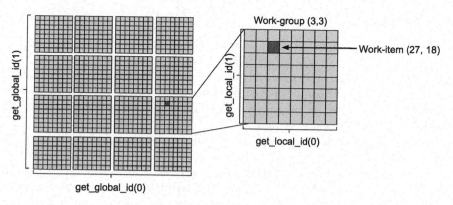

FIGURE 3.3: Example of thread organization in OpenCL. (© Juan Fumero 2017.)

Kernel functions: Kernels are the functions that are executed on a target device. OpenCL defines a set of new modifiers and tokens that are used as extensions of C99 functions. Listing 3 shows an example of how to perform vector addition in OpenCL. OpenCL adds the modifiers `kernel` to indicate that this function is the main kernel and `global` to indicate that data is located on the global memory of the target device.

Listing 3—Vector addition in OpenCL

```
1  kernel vectorAdd(global float*a, global float*b, global float*c) {
2      int idx = get_global_id(0);
3      c[idx] = a[idx] + b[idx];
4  }
```

Note that there is no loop that iterates over the data for the code shown. This is due to the fact that OpenCL maps an OpenCL kernel into a N-dimensional index space. Figure 3.3 shows an example of a 2D kernel representation in OpenCL. OpenCL can execute kernels of 1D, 2D, and 3D. Following the example in Figure 3.3, the index space is organized in a 2D block. Each block within a work-group contains a set of work-items (threads). Each work-item indexes data from the input data set using its coordinates (x, y)—if it is a 2D block. The OpenCL execution model provides built-in support for accessing and locating work-items, either using the global identifiers (for instance, the `get_global_id(0)` accesses a work-item in the first dimension), or using the functions to locate a work-item within a work-group (for instance, as shown in the right side of Figure 3.3). One of the key aspects of this organization is that threads in different work-groups cannot share memory or add barriers to wait for other groups of threads in another work-group.

Programming Model

A host program en-queues the data and the function to be executed on the target device. An OpenCL kernel follows the SIMT (Single Instruction Multiple Thread) model, in which each thread maps an item from the input data sets and performs the computation. The host issues the kernel and a bunch of threads to be executed on the target device. Then, this kernel is first submitted to the target device and executed.

If the target device is a GPU, it organizes the threads in blocks of 32 threads (or 64 threads, depending on the vendor and the GPU architecture) called `warp` or `wave-fronts`. Those threads are issued to a specific CU on the GPU. Each CU (or SM using the NVIDIA terminology) also contains a set of thread schedulers that issues an instruction per cycle. Those threads are finally located on the GPU physical cores. Threads within the same CU can share memory.

Memory Model

The OpenCL standard also defines the memory hierarchy and memory objects to be used. Figure 3.4 shows a representation of the memory hierarchy in OpenCL. The bottom shows the host memory, the main memory on the CPU. The top of the figure shows a compute device (e.g., a GPU) attached to a host. The target device also contains a region for global memory, in which input/output buffers are copied to/from. All work-groups can read and write into the device global memory. Each work-group contains its own local memory (shared memory using the NVIDIA terminology). This memory is much faster than the global memory but very limited in space. All work-items within a work-group can read and write into its own local memory. In its turn, work-items can access private memory (register files of the target device). Data in private registers is much faster to access than data in local memory. OpenCL programmers have fully control over all of these memory regions.

3.2.2.1 CUDA

Common Unified Device Architecture (CUDA) [118] is an explicit parallel programming framework and a parallel programming language that extends C/C++ programs to run on NVIDIA GPUs. CUDA shares many principles, terminology, and ideas with OpenCL. In fact, OpenCL was primarily inspired from the CUDA programming language. The difference is that CUDA is slightly simpler and more optimized due to the fact that it can only execute on NVIDIA hardware. This section shows an overview of the CUDA architecture and programming model.

CUDA Architecture

Figure 3.5 shows a representation of the CUDA architecture. The bottom of the stack shows a GPU with CUDA support (which are the majority of the

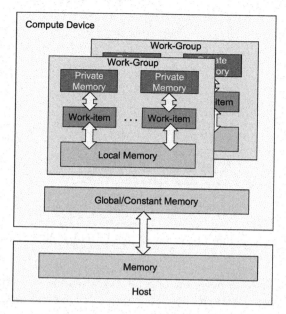

FIGURE 3.4: OpenCL memory hierarchy. (© Juan Fumero 2017.)

NVIDIA GPUs). Then, a driver at the operating system (OS) level to compile and run CUDA programs, and the CUDA driver, which performs low-level tasks and generates an intermediate representation called PTX (Parallel Thread Execution). The CUDA driver also provides an Applications Programming Interface (API) in which programmers can directly make use for developing applications for GPUs. This API is very low-level and it is very similar, in concept, to the OpenCL API and programming model.

The CUDA architecture also provides a CUDA runtime (CUDART) that facilities programming through language integration, compilation, and linking of user kernels. The CUDA runtime automatically performs tasks such as device initialization, device context initialization, loading the corresponding low-level modules, etc.

CUDA also exposes to developers a set of libraries, such as CUDA Trust for parallel programming, cuBLAS for scientific computing, and cuDNN (CUDA accelerated version for Deep Neural Networks) among many others.

CUDA Programming Overview

CUDA programs are executed on systems with GPUs attached via PCI-express to the main CPU. In the CUDA terminology, as well as in OpenCL, a host means the main director where the GPU execution is organized. A device is the target GPU in which a set of kernels is launched and executed for acceleration.

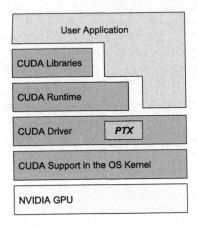

FIGURE 3.5: Overview of the CUDA architecture.

Figure 3.6 shows an overview of a CUDA program. A CUDA application is composed of two parts: the host code, which is the code to be executed on the main CPU, and the device code, which is the function to be executed and accelerated on the GPU. In a similar way to OpenCL, CUDA programs make use of streams to communicate commands between the host and devices. This architecture is practically identical to an OpenCL program's view explained in the previous section.

The main CPU contains its own physical memory. Besides that, the GPU also contains its own physical memory. The typical workflow for a CUDA application is as follows:

1. Allocate the host buffers.

2. Allocate the device buffers.

3. The CPU issues a data transfer from the host to the device (H→D) to copy the data on the device.

4. The CPU issues to launch a CUDA kernel. It sets the number of CUDA threads.

5. The CPU issues a data transfer from the device to the host (D→H) to copy the results to the main CPU.

CUDA Programming

Given the CUDA architecture, developers can write applications using three different levels of abstraction: the CUDA library, the CUDA runtime API, and the CUDA driver API. Each level of abstraction within the CUDA architecture provides an API for developing GPU applications. Many programmers prefer using libraries, mostly at the beginning of development. Additionally, the CUDA runtime API extends the C99 language (as OpenCL) with new

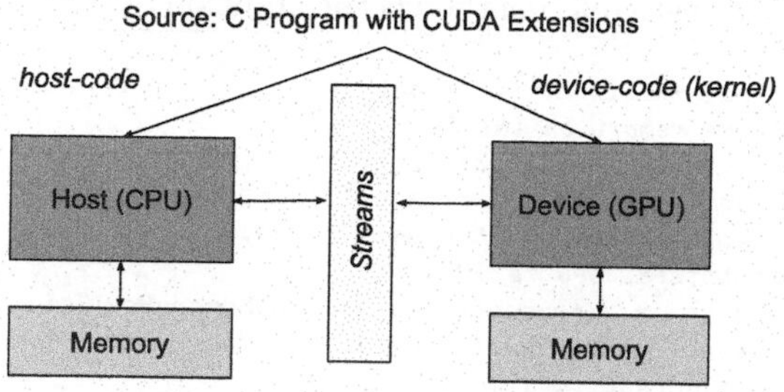

FIGURE 3.6: Overview of a CUDA program.

modifiers and tokens that express parallelism for CUDA. A CUDA driver API provides more control and tuning of CUDA applications within the physical architecture.

Listing 4 shows an example of using CUDA for vector addition. This kernel (CUDA function to be executed on the GPU) is very similar to the vector addition in OpenCL shown in Listing 3. The new language token __global__ tells the CUDA compiler that the following function corresponds to the GPU code. The code inside the function corresponds to the sequential C code for computing the vector addition for a single thread. CUDA, as well as OpenCL, uses the thread-id to index the data in an index-space range configuration. In this example, the kernel obtains its thread-id by querying the index within a 1D configuration. To query the thread-id, CUDA provides a set of high-level built-ins such as `threadIDx.x`. Then, the kernel uses this thread-id to obtain the data-item at that given location and performs the computation.

Listing 4—CUDA kernel for vector addition

```
1 __global__ void addgpu(float *a, float *b, float *c) {
2     int tid = blockIdx.x;
3     c[tid] = a[tid] * b[tid];
4 }
```

Listing 5 shows the host-code to execute the vector addition using the CUDA runtime API. It first allocates the data on the GPU by using the function `cudaMalloc`. Then, it copies the data from the host to the device using the `cudaMemcpy` function. Then, the host launches the CUDA kernel using a C extension to indicate to the CUDA compiler to execute a CUDA kernel. The CUDA kernel uses two parameters between the symbols <<<,>>> to indicate the number of threads to run and the number of blocks to run. CUDA threads, as well as in OpenCL, are organized in blocks. Therefore, the user controls the amount of threads per block and the number of blocks to use. Finally, once the CUDA kernel has been executed, the host issues a copy back from the device to the host to obtain the result (line 12).

Listing 5—CUDA host code for the vector addition

```
1  void compute(float *a, float *b, float  *c) {
2     // Allocate the memory on the GPU
3     cudaMalloc((void **)&dev_a, N*sizeof(float));
4     cudaMalloc((void **)&dev_b, N*sizeof(float));
5     cudaMalloc((void **)&dev_c, N*sizeof(float));
6     // Memory transfers: CPU to GPU
7     cudaMemcpy(dev_a, a, N*sizeof(float), cudaMemcpyHostToDevice);
8     cudaMemcpy(dev_b, b, N*sizeof(float), cudaMemcpyHostToDevice);
9     // Launch CUDA Kernel into GPU
10    addgpu<<< N, 1 >>>(dev_a, dev_b, dev_c);
11    // Memory transfers: GPU to CPU
12    cudaMemcpy(c, dev_c, N*sizeof(int), cudaMemcpyDeviceToHost);
13 }
```

The CUDA example just shown corresponds to the use of the CUDA runtime API. CUDART takes responsibility for querying all devices available, initializing them, creating the context in which sends commands, creating the stream (objects to send commands to the GPU), compiling the CUDA kernel to PTX, and running it. GPU and CUDA experts could also use the CUDA driver API, in which all these operations are fully controlled by the programmer. Normally, CUDA driver is also used by optimizing compilers and libraries, in which CUDA code can be auto-generated.

CUDA or OpenCL?

CUDA and OpenCL are very similar. Both use the same programming model and the same core concepts. Furthermore, both require deep-level expertise from the programmers' perspective. A good understanding of the GPU architecture in combination with the programming model are crucial to understand and implement applications with such models.

However, CUDA is much simpler to start with because it simplifies its model to the NVIDIA GPUs. Furthermore, CUDA is said to obtain much higher performance due to the fact that it is specially optimized for NVIDIA GPUs. Additionally, CUDA offers different levels of abstraction, such as high-level libraries, CUDA syntax, and the CUDA driver. This makes CUDA programming more flexible and suitable for many industries and academics from many different fields.

Nonetheless, OpenCL is more portable and can execute code on any OpenCL-compatible device that currently includes CPUs, GPUs from various vendors, and FPGAs (e.g., from Xilinx and Intel Altera).

3.2.3 Intermediate Representations for Heterogeneous Programming

An Intermediate Representation (IR) is a low-level language that is used by compilers and tools to optimize the code. Compilers first translate high-level source code to an intermediate representation and then optimize this IR performing several passes that implement a number of different analyses and optimizations. In the end, the resulting IR is translated into efficient machine code. This section presents a number of IRs specifically designed for heterogeneous architectures.

HSA

Heterogeneous System Architecture [158] is a specification developed by the HSA foundation for cross-vendor architectures that specifies the design of a computer platform that integrates CPU and GPUs. This specification assumes a shared memory system that connects all the computing devices. HSA is optimized for reducing the latency of data transfers across devices over the shared memory system.

The HSA foundation also defines an intermediate language named HSAIL (HSA Intermediate Language) [159]. The novelty of this IR is that it includes support for exceptions, virtual functions, and system calls. Therefore, GPUs that support HSAIL enable applications running on them to benefit from these higher-level features.

Taking the ideas developed in the HSA and HSAIL, AMD developed ROCm [160], a heterogeneous computing platform that integrates CPU and GPU management into a single Linux driver. ROCm is directly programmed using C and C++. The novelty of ROCm is that there is a single CPU and GPU compiler that targets both types of devices for C++, OpenCL, and CUDA programs.

CUDA PTX

CUDA Parallel Thread Execution (PTX) is an intermediate representation developed by NVIDIA for CUDA programs. CUDA, as shown in the previous section, is a language based on C99 for expressing parallelism. Then, a CUDA compiler (e.g., `nvcc`), translates the CUDA code to the PTX IR. At runtime, an NVIDIA driver finally compiles the PTX IR into efficient GPU code. CUDA PTX, as well as CUDA, is only available for NVIDIA GPUs.

Listing 6 shows an example of CUDA PTX for the vector addition computation. This IR is similar to an assembly code for GPUs. First, it obtains the parameters that are passed as arguments (`ld.param`). Then, it obtains the respective addresses from global memory using the `cvta.to.global` instruction, obtains the thread-id, and sums the elements. The result is finally stored using the `st.global` instruction.

SPIR

The Standard Portable Intermediate Representation (SPIR) [231] is an LLVM IR-based binary intermediate language for compute graphics and compute kernels. SPIR was originally designed by the Khronos Group in order to be

---

Listing 6—Example of CUDA PTX for a vector addition

```
1  ld.param.u64           %rd1,   [_Z5helloPcPi_param_0];
2  ld.param.u64           %rd2,   [_Z5helloPcPi_param_1];
3  cvta.to.global.u64     %rd3,   %rd1;
4  cvta.to.global.u64     %rd4,   %rd2;
5  mov.u32                %r1,    %tid.x;
6  cvt.u64.u32            %rd5,   %r1;
7  mul.wide.u32           %rd6,   %r1,    4;
8  add.s64                %rd7,   %rd4,   %rd6;
9  ld.global.u32          %r2,    [%rd7];
10 add.s32                %r3,    %r2,    %r2;
11 add.s64                %rd8,   %rd3,   %rd5;
12 st.global.u8           [%rd8], %r3;
13 ret;
```

---

used in combination with OpenCL kernels. Instead of directly compiling the source code of OpenCL kernels to machine code, compilers can produce code in the SPIR binary format. The main reason to represent SPIR in binary format is to circumvent possible license issues because of the source code distribution of the compute kernels on different devices. SPIR is currently used by optimizing compilers and libraries to distribute heterogeneous code across multiple heterogeneous architectures. Section 3.5.1.1 provides more details about SPIR.

3.2.4   Hardware Synthesis: FPGA High-Level, Synthesis Tools

High-Level Synthesis (HLS) tools enable the programming of FPGAs in a higher level of abstraction with high-level languages (HLLs). Raising the abstraction level and reducing the long design cycles, makes FPGAs more accessible and easier to adopt in modern computing systems, like data centers. The traditional approach for programming an FPGA was initially limited to hardware description languages (HDLs), such as VHDL and Verilog. However, using HDLs is a tedious process, and it requires high expertise and deep understanding of the underlying hardware. This results also in extensive periods of programming and long design cycles.

Figure 3.7 provides a high-level overview of the workflow in most HLS tools. User starts by providing a design specification written in C/C++ or SystemC along with a number of directives (also called **pragmas**). These directives allow users to provide hints to HLS compilers; for instance, a program may contain **for** loops without any dependencies that can be unrolled or pipelined. These directives guide the compiler to map a number of optimizations, such as loop unrolling, loop pipelining, or memory partitioning into the hardware. Then the initial program specification is compiled, and a formal model is produced. The information acquired during the compilation process allows the HLS to

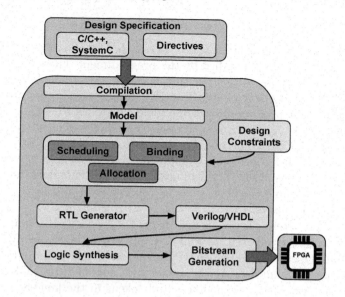

FIGURE 3.7: Overview of HLS tools workflow.

define the type of resources needed (memory blocks, LUTs, buses, etc.), and then to schedule these operations into cycles. When all these processes come to an end, and the design constraints (e.g., timing, power, and area) are met, the RTL generator provides a synthesizable model of the hardware design for the input program. Finally, logic synthesis generates a hardware configuration file which contains a hardware implementation of the initial design specification.

There is a large selection of HLS tools. The most popular and widely adopted ones at the time of writing are the Vivado HLS [31] by Xilinx, and Intel FPGA HLS [28]. Both take as input C/C++ programs and parallel code, such as OpenCL, and produce the corresponding VHDL/Verilog programs along with the machine code in binary form for the FPGA device. Other examples of HLS tools are the LegUp [87], MaxCompiler [29], Bambu HLS [312], Bluespec [26], and Catapult C [27].

3.3   Heterogeneous Programming Languages

This section describes how modern programming languages integrate with the models described previously to enable heterogeneous execution. We classify those languages in (i) unmanaged programming languages, such as C/C++ and Fortran, (ii) managed languages, such as Java, Python, Javascript, and Ruby, and (iii) high-level domain-specific programming languages, such as R.

3.3.1　Unmanaged Programming Languages

Unmanaged programming languages is a term that refers to those languages in which memory is directly handled by the programmers. These languages tend to be lower-level compared to those that automatically manage memory (e.g., Java). **C/C++** and **Fortran** are the most common unmanaged programming languages used for programming heterogeneous architectures such as GPUs, CPUs and FPGAs. In fact, many standards for heterogeneous programming are designed for C and Fortran. Furthermore, the majority of applications developed for High-Performance Computing (HPC), such as physics simulation, particle physics, chemistry, biology, weather forecast, and more recently, big data and machine learning, tend to use low-level and high-performance programming languages.

To program heterogeneous devices from C/C++ and Fortran, developers can use both directive-based, as well as explicit heterogeneous programming models. Both OpenACC and OpenMP include all directives for both languages in the standard definition. At the same time, OpenCL and CUDA extend the C99 standard with new modifiers and tokens in the language to identify parallelism.

The advantage of programming in C or Fortran using any of the standards described in the previous sections is that all rules, syntax, execution models, and memory models are well defined for the target language. On the contrary, the disadvantage of using unmanaged languages is that they are low-level and require higher expertise from the programmers. Higher-level programming languages allow for faster prototyping, and thus quicker development cycles, while requiring less expertise from the programmers, at the cost of reduced control over the execution of the program.

3.3.2　Managed and Dynamic Programming Languages

Managed programming languages is a term that refers to those languages whose memory is automatically managed by a runtime system provided by the language implementation. Some examples are languages like Java and C#, in which objects are allocated in a managed heap area. The implementation of those languages includes a language Virtual Machine (VM) that executes the applications and automatically manages memory, including object de-allocation. This removes the burden of memory management from the programmer, along with the potential accompanying bugs, allowing them to focus on other aspects of their application.

Furthermore, some managed programming languages are un-typed, like Python, Ruby and Javascript. From the programmer's perspective, this makes application development much easier and faster, allowing programmers to focus on what they want to solve instead of how to efficiently implement it. Those programming languages rely on an efficient language VM, an interpreter and a compiler to execute efficiently the input programs.

TABLE 3.1: Raking of the 10 most popular programming languages using three different sources: TIOBE Index, IEEE, and PYPL.

Language Rank	TIOBE	IEEE	PYPL
1	Java	Python	Python
2	C	C++	Java
3	C++	Java	Javascript
4	Python	C	C#
5	VBasic	C#	PHP
6	C#	PHP	C/C++
7	PHP	R	R
8	Javascript	Javascript	Objective-C
9	SQL	Go	Swift
10	Swift	Assembly	Matlab

Table 3.1 shows a ranking of the most popular programming languages in October 2018 using three different sources: TIOBE Index[2], IEEE[3], and PYPL[4]. Each source reference uses a different search criteria for ordering the programming languages by its popularity, such as Github projects, Google searches, and academic papers. What all have in common is that the majority of the languages in the top 10 are managed-programming languages, such as Java, C#, Python, and Javascript.

The question that arises after looking at the data from Table 3.1 is how to use accelerators that follow the programming models listed in the previous section from managed languages. There are currently two main techniques to program GPUs and FPGAs from managed languages:

Using external libraries: accelerated-libraries are implemented directly in C/C++ and Fortran. They provide a set of common operations, normally operations for array and matrix computation that are exposed to the managed language through a language interface from the C level. For example, in the case of Java, they can be exposed using the Java Native Interface (JNI), in which operations are implemented using C code and called from the Java side.

Using a wrapper: the use of accelerators is directly programmed from the high-level programming languages using the same set of standards API calls. This provides fine-tune control over the applications to be executed on the accelerators, but contradicts with the nature of the managed programming languages since it requires low-level understanding.

---

[2]https://www.tiobe.com/tiobe-index/
[3]https://spectrum.ieee.org/static/interactive-the-top-programming-languages-2018
[4]http://pypl.github.io/PYPL.html

Examples of external libraries are ArrayFire [264] and PyTorch[5]. ArrayFire is a library for programming parallel CPUs and GPUs in C, C++, Java, and Python. It is designed for matrix computation using CUDA and OpenCL underneath, and it contains numerous highly-optimized-function GPUs for signal processing, statistics, and image processing. PyTorch is a Python library for deep learning on GPUs. It exposes a set of high-level operations that are efficiently implemented in CUDA.

Some example of wrappers are JOCL [217], for programming OpenCL within Java, and JCUDA [386], for programming CUDA within Java. Both expose a set of native calls to be invoked from Java that matches the OpenCL and CUDA definitions respectively. Kernels are expressed as a Java string that is then passed to the compiled and execute via JNI calls.

Challenges

Although high-level and managed-programming languages ease the development process, the higher the abstraction the more difficult it becomes to optimize the code for the target platform. For example, in Java, all arrays are copied from Java's side to the JNI's side. This means that JNI makes an extra copy for all arrays, and therefore, the performance of applications will be different compared to those implemented in unmanaged languages like C. Programming heterogeneous devices from managed languages increases the complexity of efficiently managing memory and types. For example, neither OpenCL not CUDA support objects, while in object-oriented high-level programming languages everything is an object. Therefore, extra effort is required by the runtime system to convert the data from their high-level representation to a low-level representation and vice versa. This process is called *marshalling/unmarshalling* and can be time-consuming depending on the objects at hand.

Furthermore, the semantics of high-level programming languages do not always match those of the heterogeneous programming models. For example, the OpenCL standard does not clarify what happens with runtime exceptions, such as a division by zero. In the case of Java, the JVM is required to throw an `ArithmeticException`, but hardware exceptions are not currently supported in CUDA or OpenCL.

3.3.3　Domain Specific Languages

Domain Specific Languages (DSLs) are languages designed for an specific purpose. For example R [205] is a programming languages specifically designed for statistics. R, as shown in Table 3.1, is one of the most popular, and has been increasing its popularity during the last three years. Many scientists use R to process big data, provide statistics, and even predict future events using machine learning due the fact that it provides an enormous quantity of external

---

[5]https://pytorch.org

modules. These external modules are normally written in lower programming languages such as C and Fortran for better performance.

In DSLs, heterogeneous devices can be programmed using external libraries and wrappers, as in managed programming languages. In the case of DSLs, however, the challenges of programming heterogeneous systems are even more complex. DSLs are mainly used because of their simplicity and their specialization in a specific domain. Introducing wrappers raises the complexity of DSLs and breaks their specialization. As a result, programmers are required to mix and understand different programming and architecture models, totally unrelated to their domain.

3.4 Heterogeneous Device Selection

This section describes how developers can select the most suitable device for executing their code. More specifically, it presents and discusses various techniques for device selection such as: offline source code analysis, machine learning models, and profile-guided selection.

3.4.1 Schedulers

In typical computing environments, such as Cloud and HPC datacenters, the infrastructure is being managed by software systems called resource schedulers. Schedulers are responsible for allocating tasks to the underlying infrastructure that can consist of clusters of computing nodes, storage, networking, etc. Resource schedulers typically operate in a master-slave manner, where a central machine (called the "master") is used to consolidate and manage the amount of different workers (i.e., slaves) by utilizing a centralized resource registry. The workers use different approaches to register themselves with the master, such as heartbeat messages, etc, following the same architecture with typical legacy "batch" schedulers such as Condor [362] and PBS-Torque [196, 89].

Modern schedulers are starting to embrace the heterogeneity of resources encountered in Cloud infrastructures. The scheduling techniques can be categorized in workload partitioning methods (static vs. dynamic), subtask-based scheduling, pipe-lining, and MapReduce-based [276].

Apache Mesos [198] uses a two-level scheduling mechanism where resource offers are made to frameworks (apps like web servers, map-reduce programs, NoSQL databases, etc. that run on top of Mesos). The Mesos master node decides how many resources to offer each framework, while each framework determines the resources it accepts and what application to execute on those resources. Apache YARN [372] is the de-facto scheduler used in modern deployments of the Hadoop ecosystem software. It also utilizes a negotiation framework where resources are being requested and provided upon requests from different applications. Both schedulers offer the possibility to "label" different execution nodes with user defined-labels, leaving the user to handle heterogeneity.

ORiON [391] and IReS [133] schedulers both can schedule heterogeneous workloads on big data clusters. ORiON can adaptively decide the appropriate framework along with the cluster configuration in terms of software and hardware resources to execute an incoming analytics job. It is based on a combination of a decision-tree-like machine learning process for resource prediction and a integer linear problem formulation for resource optimization. IReS can identify and materialize the optimal sequence of analytics engines required to execute a data-processing pipeline in an abstract execution workflow given by the user by utilizing a dynamic-programming technique. Although both schedulers can work with heterogeneous software systems, the hardware heterogeneity is not exploited.

TetriSched [369] also schedules heterogeneous resources by forming a Mixed Integer Linear Programming problem, nevertheless it is not aware of any task relations in the submitted workload (i.e., it is DAG-oblivious), similar to ORiON.

In [378] the authors present a scheduling methodology for tasks that can be executed on both CPU and GPU devices. They evaluate their methodology with different complex algorithms (i.e., kernels) such as bfs, BlackScholes, Dotproduct, and QuasirandomG, and they show that they can predict and deploy their kernels to the most appropriate engine. Their approach is based on predictive modeling, i.e., on identifying the important kernel features that affect the performance (i.e., execution time). They utilize these features to build a machine learning model based on support vector machines and they use it to predict the execution time upon workload arrival. Nevertheless, they are also DAG-oblivious and they do not explore different performance metrics.

A similar approach is also followed by [377], where a machine learning model is built to predict execution time based on carefully selected features. Their main differentiation compared to [378] is that they study the effect of concurrently executing kernels on the same device (i.e., merging) to identify whether a speedup can be achieved or not.

3.4.2   Platforms

Over the last years many distributed and parallel data analytics frameworks have emerged. These systems process in parallel large amounts of data in a batch or streaming manner. In this section, we investigate to what extent the heterogeneity of modern datacenters is exploited to improve application performance.

GFlink [103] is a distributed stream-processing platform that extends Apache Flink[93]. Its key-feature is that, apart from CPU processors, it can also execute tasks on top of GPU accelerators. To achieve that, it builds upon the master-slave architecture of Flink, and introduces processes (GPUManagers) that are responsible for the GPU management. Execution is coordinated by a single master which schedules tasks to workers either for CPU or GPU execution. The scheduling scheme is locality-aware and manages to avoid unnecessary communication, while it achieves load-balancing among GPUs. Nevertheless, details on the scheduling algorithm are not provided.

GFlink runs in the Java Virtual Machine (JVM). The standard way of communication between JVM and the GPU devices is to serialize memory objects that live in the JVM heap, and transfer data over the PCIe bus. However, as this process incurs high serialization-deserialization cost, GFlink opts for a more efficient memory management: Users can define custom structures that live off-heap and operate directly on them. This way, object serialization is avoided. Moreover, the master can instruct the GPU devices to cache data and reduce the time spent in copying operations.

In [102] the authors propose EML [129], another system that combines Flink and GPUs to proccess large datasets. Although very similar to GFlink, EML does not use a separate process for managing GPU execution but modifies the existing Flink TaskManagers to support both execution modes.

Spark-GPU [389] is a CPU-GPU hybrid data analytics system based on Apache Spark[393]. Data in Spark is modeled as RDDs [392] and consumed in a one-element-at-a-time fashion. However, this does not match the massive parallelization a GPU can offer and leads to resource underutilization. Spark-GPU overcomes this issue by extending Spark's iterator model and introducing GPU-RDD: a new structure which buffers all its data in native memory and can be consumed in a per-element or per-block basis. Moreover, as GFlink, Spark-GPU utilizes native memory instead of the Java heap space in order to avoid the excessive cost of often serialization-deserialization tasks.

For taking advantage of the capabilities of a GPU, isolation should exist and only one application should run on the device at a given time. At the moment when Spark-GPU was developed, resource managers like YARN [372] and Mesos [198] did not offer isolated execution for GPUs. Thus, to fully exploit the potential of the underlying hardware, Spark-GPU comes with its own custom resource manager.

Regarding task scheduling, Spark-GPU extends the Spark-SQL query optimizer and creates a GPU-aware version of it. A rule-based optimizer adaptively schedules user queries to the most beneficial hardware device. The selection of a GPU for a task depends on whether an algorithm fits the GPU execution model.

HeteroSpark [250] is another GPU-accelerated Spark-based architecture. Applications that run on top of HeteroSpark can explicitly choose whether or not a task should be executed on a GPU device.

Contrary to the aforementioned systems, objects are serialized and deserialized on demand. Furthermore, accelerating a Java application with HeteroSpark does not happen in an automated way. It requires the following steps: (i) wite a GPU kernel, (ii) develop a wrapper in C that makes use of the Java Native Interface (JNI) in order to create a Java API for the kernel, and (iii) deploy in Spark.

SWAT [184] and SparkCL [339] are two open-source frameworks that are able to accelerate user-defined Spark kernels by using OpenCL. While SWAT supports only GPU acceleration, SparkCL targets a broader range of processing devices, like GPUs, APUs, CPUs, FPGAs and DSPs.

Both systems use the Aparapi [47] framework for translating Java methods to OpenCL kernels and the communication with the accelerators is based on the on-demand serialization-deserialization of Java objects.

A flurry of activity in the development of heterogeneous systems also exists in the realm of machine learning. The well-known Google's TensorFlow [34] is a system that enables the training of machine learning algorithms over large-scale heterogeneous environments. The devices TensorFlow supports are: CPUs, GPUs, and TPUs (Tensor Processing Units, a unit specialized for machine learning applications). The execution model is based on a dataflow graph, where nodes represent tasks and edges data dependencies. Each task can run on a different kind of processor, and the user can explicitly select the device of preference for each separate task. In case she does not, Tensorflow may employ an automatic placement algorithm. However, the algorithm takes into account only basic considerations (e.g., a stateful operation and its state should be placed in the same device) and is not yet mature enough to take decisions that guarantee optimal performance in large-scale clusters.

Tensors, the logical abstraction of Tensorflow's data structures, consist of primitive values that can be efficiently interpreted by all supported devices. Thus, concerns about the overhead of serialization-deserialization do not apply in this case.

3.4.3   Intelligence

The approaches presented in this subsection alleviate the burden of manually selecting the most beneficial mapping between application tasks and heterogeneous hardware. Contrarily, they automatically determine the processing element that best fits each application, input, and configuration, making educated decisions on the preferred computing resource based on intelligence derived from code analysis, offline profiling, and machine learning techniques.

Qilin [259] is a heterogeneous programming system that relies on an adaptive mapping technique, which automatically maps computations to heterogeneous devices. Qilin offers a programming API built on top of C/C++, which provides primitives to express parallelizable operations. The Qilin API calls are dynamically translated into native CPU or GPU code though the Qilin compiler. More specifically, the compiler first builds a Directed Acyclic Graph (DAG), where nodes represent computational tasks of the application and edges represent data dependencies. Then, it automatically finds the near-optimal mapping from computations to processing units relying on a per-task linear regression model that provides execution time estimations for the current problem size and system configuration. The regression model of each task is trained on-the-fly, during the task's first execution under Qilin: The input is divided in multiple parts and assigned equally to the CPU and the GPU. Execution-time measurements are fed to curve-fitting techniques to construct linear equations, which are then used as projections for the actual execution time of the task over CPU or GPU.

Hayashi et al. [195] have developed a system for compiling and optimizing Java 8 programs for GPU execution extending IBM's just-in-time (JIT) compiler. One of the extensions consists in adding the capability of automatic CPU vs. GPU selection. This is achieved by using performance heuristics that rely on supervised machine learning models, i.e., a binary Support Vector Machine (SVM) classifier constructed in an offline manner using as training data measurements of actual program executions over various input datasets. The classifier input dimensions consist of a set of static code features that affect performance, extracted at compile time: loop range of a parallel loop, number of instructions per iteration, number of array accesses and data transfer size. The output is the preferred computing resource (CPU vs. GPU) that optimizes the program performance.

The work in [181] proposes an approach to partitioning data-parallel OpenCL tasks among the available processing units of heterogeneous CPU-GPU systems. Code analysis is used during compilation to extract 13 static code features, which include information such as the amount of int and float operations, the number of memory accesses, the size of data transferred, etc. Principal Component Analysis (PCA) is applied to reduce the dimensionality of the feature space and normalize the data. The normalized, low-dimensional data are then passed through a two-level machine-learning predictor that performs hierarchical classification to determine the optimal partitioning for the corresponding OpenCL program: The first level distinguishes CPU- and GPU-only optimal executions using a binary Support Vector Machine (SVM) classifier, while the second one handles the cases where the best performance is achieved when distributing execution over both GPU and CPU. The latter is performed by classifying programs to nine different categories along the spectrum between CPU- and GPU-only (i.e., 10% CPU-90% GPU, 20% CPU-80% GPU, etc.) using again SVM models. Both models are trained in an offline manner with profiling measurements over various partitioning schemes.

HCl (Heterogeneous Cluster) [227] is a scheduler that maps heterogeneous applications to heterogeneous clusters. HCl represents heterogeneous applications as Directed Acyclic Graphs (DAG), where nodes stand for computations and edges represent data transfer between connected nodes. Taking into account the I/O volume between DAG tasks, the available hardware resources and the runtime estimations of each task on each resource, HCl exhaustively evaluates all possible combinations of task to node mappings and selects the the global optimum, i.e., the execution schedule that optimizes the entire task graph rather than each task separately.

DeepTune [119] is an optimization framework that relies on machine learning over raw code. Its goal is to bypass the code feature selection stage involved in the techniques presented so far. Feature selection requires manual work by domain experts and heavily affects the quality of the resulting machine learning model. One of the demonstrated use cases of DeepTune is the creation of a heuristic to select the optimal execution device (CPU or GPU) for an OpenCL kernel.

The architecture of DeepTune is a machine-learning pipeline: after the source code is automatically rewritten according to a consistent code style, it is transformed into a sequence of integers using a language-specific vocabulary which maps source code tokens to integer indices. Using embeddings that translate each token of the vocabulary to a low-dimensional vector space, the sequence of integers is transformed into a sequence of embedding vectors that capture the semantic relationship between tokens. A Long Short-Term Memory (LSTM) neural network is then used to extract a single, fixed-size vector that characterizes the entire sequence of embedding vectors. During the last stage, the resulting vectors, i.e., the learned representations of the source code is fed to a fully connected, two-layer neural network to make the final optimization prediction: The first layer has a constant number of neurons, while the second layer consists of one neuron per possible heuristic decision.

3.5   Emerging Programming Models and Architectures

This section describes the state-of-the-art and current developments in the areas of heterogeneous programming models and architectures. The overall description is divided into two parts. The first part describes the most common hardware-oriented solutions, where hardware heterogeneity is tackled by deploying low-level software drivers and end-to-end frameworks which unify the different sets of underlying hardware. The second part describes a different approach in which an abstraction layer is generated, mostly using contemporary containerization techniques, that handles all different hardware in a seamless matter. The user code thus remains hardware-agnostic but is executed in a highly efficient manner.

3.5.1   Hardware-bound Collaborative Initiatives

3.5.1.1   Khronos Group

The Khronos Group[6] is probably the leading industry consortium for the creation of open-standard, royalty-free application programming interfaces (APIs) for authorization and hardware-accelerated playback of graphics and dynamic media in general, onto a large variety of devices. The activity of the consortium is mainly towards video, however, all heterogeneous platforms benefit from software and APIs which enhance the abilities of the underlying

---

[6]https://www.khronos.org

hardware. In addition, the consortium is also interested in parallel computation efficiency thus having several of its solutions addressing related issues.

Vulkan

Vulkan[7] is a cross-platform graphics and compute API which provides highly-efficient access to all modern GPU hardware resources. Its prime scope is to offer higher performance in 3D graphics applications, such as video and interactive media, with a more balanced CPU/GPU utilization. More specifically, the API ensures that the GPU only executes shaders while, the CPU executes everything else. In addition to its lower CPU usage, Vulkan is also able to better distribute work among multiple CPU cores, offering a reduced driver overhead, and extensively utilizing batching thus releasing more computational cycles for the CPU. The latest version of the API, Vulkan 1.1 also supports subgroup operations, an important new feature since it enables highly-efficient sharing and manipulation of data between multiple tasks running in parallel on a GPU.

SPIR

Standard Portable Intermediate Representation (SPIR)[8] is an intermediate language for parallel compute and graphics, originally developed for use with OpenCL. SPIR has now evolved into a cross-API intermediate language that is fully defined by Khronos with native support for shader and kernel features used by affiliated APIs. The current version, SPIR-V 1.3 was released on March 7th, 2018 to accompany the launch of Vulkan 1.1, and is designed to expand the capabilities of the Vulkan shader intermediate representation by also supporting subgroup operations thus enabling enhanced compiler optimizations. SPIR-V is the first open-standard, cross-API intermediate language for natively representing parallel compute and graphics. It is part of the core specifications of OpenCL 2.1, OpenCL 2.2, and it is supported in an OpenGL 4.6 extension but is no longer using LLVM.

However, Khronos has open-sourced SPIR-V/LLVM conversion tools to enable construction of flexible toolchains that use both intermediate languages. SPIR-V is catalyzing a revolution in the language-compiler ecosystem—it can split the compiler chain across multiple vendors' products, enabling high-level language front-ends to emit programs in a standardized intermediate form to be ingested by Vulkan, OpenGL, or OpenCL drivers. For hardware vendors, ingesting SPIR-V eliminates the need to build a high-level language source compiler into device drivers, significantly reducing driver complexity, and will enable a broad range of language and framework front-ends to run on diverse hardware architectures.

---

[7]https://www.khronos.org/vulkan/
[8]https://www.khronos.org/spir/

SYCL

SYCL[9] is a high-level abstraction layer that builds on the underlying concepts, portability, and efficiency of OpenCL, thus allowing code for heterogeneous processors to be written in a "single-source" style using completely standard C++. SYCL single-source programming enables the host and kernel code for an entire application to be contained in the same source file, in a type-safe way, and with the simplicity of a cross-platform asynchronous task graph. SYCL includes templates and generic lambda functions to enable higher-level application software development. SYCL not only introduces the power of single-source modern C++ to the SPIR world, but with its recent 1.2.1 revision 3, it integrates features related to machine learning environment requirements. In addition, Khronos provides an open-source implementation to experiment with and provide necessary feedback[10].

OpenKODE

OpenKODE[11] is a royalty-free, open standard that combines a set of native APIs to increase source portability for rich media and graphics applications. It reduces mobile platform fragmentation by providing a cross-platform API for accessing operating system resources, and a media architecture for portable access to advanced mixed graphics acceleration. OpenKODE also includes the OpenKODE Core API that abstracts operating system resources to minimize source changes during application porting.

3.5.1.2 OpenCAPI

OpenCAPI[12] is an Open Interface Architecture that allows any microprocessor to attach to (i) Coherent user-level accelerators and I/O devices, (ii) Advanced memories accessible via read/write or user-level DMA semantics, (iii) Agnostic-processor architectures. This initiative aims to create an open, high-performance bus interface based on a new bus standard called Open Coherent Accelerator Processor Interface (OpenCAPI) and grow the ecosystem that utilizes this interface. The main drive behind this initiative is the constantly increasing acceleration on computing and advanced memory/storage solutions that have introduced significant system bottlenecks in today's current open-bus protocols, all of which require a technical solution that is openly available.

---

[9] https://www.khronos.org/sycl/
[10] https://github.com/Xilinx/triSYCL
[11] https://www.khronos.org/openkode/
[12] https://opencapi.org/

3.5.2   Serverless Frameworks

Function-as-a-Service (FaaS) approach has recently emerged and immediately gained a lot of interest in the Cloud services community. Initially, only Cloud providers provided such a functionality but recently several open-source approaches emerged. This trend was also boosted by the rise of container orchestrators simplifying the deployment and execution of such frameworks in a serverless way.

Serverless services enable the developers to compose an application from several multi-language services disengaging different teams from utilizing the same programming language. Moreover, since, every individual Cloud provider is not able to meet all of the requirements of every service composing a platform, it is common to deploy services to multiple Cloud providers according to the needs of each service. This requirement can also be covered by serverless architectures enabling the deployment of a platform in heterogeneous Cloud infrastructures.

OpenFaaS

OpenFaaS [17] is an open-source FaaS framework that utilizes Docker and Kubernetes to host serverless functions. OpenFaaS is able to package any process as a function and execute it anywhere utilizing containerization technology; it supporting many programming languages. Since OpenFaaS is open source, it can be easily extended to meet any specific requirements, for instance to add support for a programming language.

Nuclio

Nuclio [14] is another serverless framework focused on high-performance events and data processing. Nuclio provides a convenient way for the user to define any function for processing data and event streams providing integration with several heterogeneous data sources. Nuclio provides a convenient SDK to write, test, and submit function code without knowledge on the entire Nuclio architecture and source code.

Fn Project

Fn Project [7] is another open-source serverless platform utilizing Docker containers since each function submitted in Fn is executed in a Docker container. Fn provides support for a broad range of programming languages and utilizes a smart load balancer for routing traffic to functions. Through the Fn FDK (Function Development Kit) a user is able to quickly bootstrap functions in any language supported by Docker, defining input source binding models, and testing the submitted functions.

Apache OpenWhisk

Apache OpenWhisk [3] is also an open-source distributed serverless platform that is event-driven since the functions are executed in response to events. OpenWhisk also uses Docker containers to manage the infrastructure and handle scaling. Again, the user is able to define a function block that gets triggered upon the reception of an event providing integration with several sources. OpenWhisk supports several programming languages and provides a CLI for managing several aspects of an OpenWhisk instance.

Kubeless

Kubeless [10] is a Kubernetes native serverless framework designed to be deployed on top of a Kubernetes cluster, leveraging Kubernetes infrastructure management, auto-scaling, routing, monitoring and other primitives. Kubeless uses a Custom Resource Definition to be able to create functions as custom Kubernetes resources. It then runs an in-cluster controller that watches these custom resources and launches runtimes on-demand. The controller dynamically injects the functions code into the runtimes and make them available over HTTP or via a PubSub mechanism.

Fission

Fission [6] is also a framework for serverless functions on Kubernetes enabling the definition of functions in any language, and it maps them to event triggers abstracting away containers creation and management. Moreover, Fission Workflows enable the orchestration of a set of serverless functions without directly dealing with networking, message queues, etc., automating the process of building complex serverless applications that span many functions.

Funktion

Funktion [8] is an open-source event-driven lambda-style programming model designed for Kubernetes. Funktion supports several event sources and connectors including many network protocols, transports, databases, messaging systems, social networks, Cloud services, and SaaS offerings. Funktion is a serverless approach to event driven microservices and focusses on being Kubernetes- and OpenShift-native rather than a generic serverless framework.

Quebic

Quebic [19] is a FaaS framework for writing serverless functions to run on Kubernetes. Currently quebic supports only Python, Java, and NodeJS. The event-driven messaging mechanism of Quebic enables invocations from an API Gateway to function or inter-functions calls. Quebic also provides automated processes for submitted functions update or downgrade and, apart from functions, Quebic can also host event-driven microservices.

Riff

Riff [21] is another open-source serverless framework that works in any certified Kubernetes environment. Again Riff is a framework designed for running functions in response to events. Since functions in Riff are packaged as containers, they can be written in a variety of languages and provides integration with several event sources. In Riff when a software function is triggered, Kubernetes, the orchestrator of Riff, spins one container and afterwards kills it off abstracting away those operations from the developers.

Serverless Framework

Serverless Framework [22] is an open-source CLI for building and deploying serverless applications. Serverless enables Infrastructure as Code defining entire serverless applications, utilizing popular serverless technologies like AWS Lambda, with simple configuration files. Serverless framework is Cloud provider agnostic and provides a simple, intuitive CLI experience that makes it easy to develop and deploy applications to public Cloud platforms. Serverless framework supports also several programming languages and provides a robust plug-ins ecosystem and built-in support for application life-cycle management.

3.6 Ongoing European Projects

This section gives a high-level overview of ongoing European projects that perform research on programming models for heterogeneous computing and study heterogeneity programming challenges. Given the fact that in the upcoming era of exascale computing [147], where systems will be capable of a quintillion 10^{18} floating-point operations per second (FOPS), all major infrastructure is expected to be vastly heterogeneous and heavily relying on GPUs [244], we have also included the most prominent ongoing projects from the relevant EU call[13].

3.6.1 ALOHA

ALOHA[14] aims to ease the deployment of deep learning (DL) algorithms on edge-nodes. To achieve its goals the ALOHA project will develop a software development tool flow, automating among others the porting of DL tasks to heterogeneous embedded architectures, their optimized mapping, and scheduling. The ALOHA project uses two distinct platforms as test-beds. The first one is a low-power Internet of Things (IoT) platform, while the second is an

---

[13]https://cordis.europa.eu/programme/rcn/702036_en.html
[14]https://www.aloha-h2020.eu/

FPGA-based heterogeneous architecture designed to accelerate Convolutional Neural Networks (CNNs).

3.6.2   E2Data

E2Data[15] proposes an end-to-end solution for Big Data deployments that will deliver performance increases while utilizing less Cloud resources without affecting current programming norms (i.e., no code changes in the original source). E2Data will provide a new Big Data paradigm, by combining state-of-the-art software components, in order to achieve maximum resource utilization for heterogeneous Cloud deployments. The evaluation will be conducted on both high-performing x86 and low-power ARM cluster architectures representing realistic execution scenarios of real-world deployments in four resource-demanding applications from the finance, health, green buildings, and security domains.

3.6.3   EPEEC

The European joint Effort toward a Highly Productive Programming Environment for Heterogeneous Exascale Computing[16], in short EPEEC, aims to deliver a production-ready parallel programming environment that will ease the development and deployment of applications on the upcoming overwhelmingly-heterogeneous exa-scale supercomputers. The project will advance and integrate state-of-the-art components based on European technology, with the ultimate goal to provide high coding productivity, high performance, and energy awareness.

3.6.4   EXA2PRO

EXA2PRO[17] is another project that aims to deliver a programming environment that will enable efficient exploitation of exa-scale systems' heterogeneity. The EXA2PRO programming environment will support a wide range of scientific applications, provide tools for improving source code quality and integrate tools for data and memory management optimization. Furthermore, it will provide performance monitoring features, as well as fault-tolerance mechanisms.

3.6.5   EXTRA

EXTRA[18] project aims to create a new and flexible exploration platform for developing reconfigurable architectures, design tools and HPC applications with runtime reconfiguration built-in from the start.

---

[15]https://e2data.eu
[16]https://cordis.europa.eu/project/rcn/215832_en.html
[17]https://exa2pro.eu
[18]https://www.extrahpc.eu/

3.6.6 LEGaTO

LEGaTO's[19] goal is to provide a software ecosystem for Made-in-Europe heterogeneous hardware composed off CPUs, GPUs, FPGAs and FPGA-based data-flow engines (DFEs). LEGaTO will leverage task-based programming models, similar to OpenMP, to ultimately achieve a one order of magnitude increase in energy efficiency. Additionally, LEGaTO will explore ways to ensure the resilience of the software stack running on the heterogeneous hardware.

3.6.7 MANGO

MANGO[20] targets achieving extreme resource efficiency in future QoS-sensitive HPC through ambitious cross-boundary architecture exploration for performance/power/predictability (PPP) based on the definition of new-generation high-performance, power-efficient, heterogeneous architectures with native mechanisms for isolation and quality-of-service, and an innovative two-phase passive cooling system. Its disruptive approach will involve many interrelated mechanisms at various architectural levels, including heterogeneous computing cores, memory architectures, interconnects, runtime resource management, power monitoring and cooling, as well as programming models. The system architecture intends to be inherently heterogeneous as an enabler for efficiency and application-based customization, where general-purpose compute nodes (GN) are intertwined with heterogeneous acceleration nodes (HN), linked by an across-boundary homogeneous interconnect. It will provide guarantees for predictability, bandwidth, and latency for the whole HN node infrastructure, allowing dynamic adaptation to applications.

3.6.8 MONT-BLANC

MONT-BLANC[21] is a long-running project, currently in its fourth phase. MONT-BLANC's aim is to provide solutions for European energy-efficient HPC. During its previous phases, among others, the project created an ARM-based HPC cluster and proposed techniques to address the challenges of massive parallelism, heterogeneous computing, and resiliency.

3.6.9 PHANTOM

PHANTOM[22] aims to enable next-generation heterogeneous, parallel and low-power computing systems, while hiding the complexity of the underlying hardware from the programmer. The PHANTOM system comprises a hardware-agnostic software platform that will offer the means for multi-dimensional

---

[19]https://legato-project.eu/
[20]http://www.mango-project.eu
[21]https://www.montblanc-project.eu/
[22]http://www.phantom-project.org/

optimization. A multi-objective scheduler decides where in the computing continuum (e.g., Cloud, embedded systems, mobile devices, desktops, data centers), at which cross-layer system level (analog, digital, hybrid analog-digital, software) and on which heterogeneous technology (GPU, FPGA, CPU) to execute each part of an application. Additionally, it orchestrates dynamically the hardware and software components of reconfigurable hardware platforms.

3.6.10 RECIPE

RECIPE[23] (REliable power and time-ConstraInts-aware Predictive management of heterogeneous Exascale systems) will provide a hierarchical runtime resource-management infrastructure to optimise energy efficiency and minimise the occurrence of thermal hotspots. At the same time it will enforce the time constraints imposed by the applications, and ensure reliability for both time-critical and throughput-oriented computation. Apart from the runtime itself RECIPE's second work package includes a task specifically focusing on programming models.

3.6.11 TANGO

The scope of TANGO[24] is to provide the means for controlling and abstracting underlying heterogeneous hardware architectures, configurations and software systems including heterogeneous clusters, chips and programmable logic devices while developing tools to optimize various dimensions of software design and operations (energy efficiency, performance, data movement and location, cost, time-criticality, security, dependability on target architectures). The key novelty of the project is a reference architecture and its actual implementation that includes the results of the research work into different optimization areas (energy efficiency, performance, data movement and location, cost, time-criticality, security, dependability on target architectures). Moreover, TANGO integrates a programming model with built-in support for various hardware architectures including heterogeneous clusters, heterogeneous chips and programmable logic devices. In addition, TANGO creates a new cross-layer programming approach for heterogeneous parallel hardware architectures featuring automatic code generation including software and hardware modeling. Last but not least, the project provides certain mechanisms which facilitate the control of all aforementioned heterogeneous parallel infrastructures in an open-source toolbox[25].

---

[23]http://www.recipe-project.eu/
[24]http://www.tango-project.eu
[25]http://www.tango-project.eu/content/beta-tango-toolbox-released-open-source

3.6.12 VINEYARD

VINEYARD[26] aims to develop an integrated platform for energy-efficient heterogeneous data centers based on servers with programmable hardware accelerators. To increase productivity on such platforms, VINEYARD also builds a high-level programming framework for allowing end-users to seamlessly utilize such heterogeneous platforms by using typical data-center programming frameworks (e.g., Storm, Spark, etc.).

3.7 Conclusions

This chapter provided a high-level overview of current programming and architecture models for heterogeneous computing. It first showed an overview of heterogeneous programming models such as OpenACC, CUDA, and OpenCL. Then, it covered programming such systems from managed programming languages such as Java, Python, and R. It also presented the state-of-the-art research and projects regarding device selection for heterogeneous computing. Finally, it showed current initiatives and ongoing European projects working towards improving and handling programming for heterogeneous computing.

---

[26]http://vineyard-h2020.eu

4

Simplifying Parallel Programming and Execution for Distributed Heterogeneous Computing Platforms

J. Ejarque

Barcelona Supercomputing Center (BSC), Barcelona, Spain

D. Jiménez-González, C. Álvarez, and X. Martorell

Barcelona Supercomputing Center (BSC) and Universitat Politècnica de Catalunya (UPC), Barcelona, Spain

R. M. Badia

Barcelona Supercomputing Center (BSC) and Spanish National Research Council (CSIC), Barcelona, Spain

CONTENTS

With the rise of the internet and the progress of information technologies, data-processing requirements have increased exponentially. Moreover, computer systems have also evolved to fulfill this requirements resulting in more complex systems. Recent trends aim to improve computing performance by incorporating different types of computing devices such as GPUs, FPGAs and CPU architectures, which are specialized for accelerating different types of algorithms as well as reducing the energy consumption. With the rise of Cloud and Fog computing, these devices can be distributed geographically and connected through the internet by means of different types of networks.

Programming applications to take profit of distributed heterogeneous computing systems is becoming a very complex and hard task because developers have to deal with different problems and management issues. In this chapter, we present a methodology to combine different task-based programming models in order to simplify the development of complex parallel applications for distributed and heterogeneous systems and what are the benefits of this proposal compared to current options.

4.1   Introduction

Programming applications for taking profit from distributed heterogeneous computing systems is becoming a very complex and hard task. It is mainly because developers have to deal with different problems and issues. The first problem that a developer has to deal with is how to exploit the parallelism of the applications. To improve the performance in new computing systems, applications have to be distributed in different software bits which can be executed in parallel in several cores of a SMP, different devices (CPU, GPU, FPGA, etc.), or different computing nodes. However, parallelising an application may introduce overheads due to the parallel management of processes, threads, tasks, synchronizations, etc. that can be significant and vary depending on the amount of tasks and where a task can be executed. Therefore, the task decomposition strategy needs to aligned with this potential overhead: at node-level task granularity needs to be coarse grain, while for executions within a node, it can be fine grain.

Besides the task decomposition strategy, the parallel implementation itself is also a challenge for the programmers. Those have to deal with processes, threads, and different devices APIs or libraries to spawn computation in accelerators, and frameworks to perform remote executions in distributed memory systems. Therefore, they may have to implement the communication and synchronization mechanisms to transfer data between the different computing locations. This problem also appears when developers use the available accelerators in a computing node, where data have to be transferred from the main node memory to the device memory. Moreover, using different architectures will require to manage data serialization in order to be able to translate data formats from one architecture to the other.

On the other hand, in order to achieve parallel efficiency, programmers should use scheduling techniques to select the best resources to execute the different application bits. Note that different computing bits can be executed in different nodes and computing devices, but depending on the node used the execution can be more efficient than in another resource.

Finally, application resource requirements may vary based on the dynamic load of the application at runtime, depending on the input parameters or the application phase. Therefore, another desired feature is the adaptation of the infrastructure conditions to the application load and vice versa. These features have to be implemented by the user by exchanging messages with the resource providers, API.

In the literature, you can find approaches, such as APIs or programming models and frameworks, which are trying to solve some of the aforementioned challenges. However, to implement an application which benefits from heterogeneity in distributed environments, developers have to combine several of these approaches. Therefore, they require implementing some glue code to integrate them into their application code. In this chapter, we present how the StarSs approach deals with all the previous challenges in order to simplify the programming of parallel applications for distributed and heterogeneous systems. This approach consists of a hierarchical combination of the COMPSs and OmpSs task-based programming models (both belonging to the StartSs programming model family) which have been integrated to efficiently execute applications in different heterogeneous computing nodes.

The rest of the chapter is organized as follows. In Section 4.2, the StarSs approach is presented and, COMPSs and OmpSs are described as two programming models that can be used by separated or integrated to exploit heterogeneity on distributed systems. Section 4.3 introduces and compares our proposal to current parallel programming approaches for heterogeneous architectures and distributed systems. Finally, in Section 4.4, we conclude and propose guidelines for future work.

4.2 StarSs: A Task-Based Approach

StarSs is a family of task-based programming models where developers define some parts of the application as tasks indicating the direction of the data required by those tasks. Based on these annotations the programming model runtime analyzes data dependencies between the defined tasks, detecting the inherent parallelism and scheduling the tasks on the available computing resources, managing the required data transfers and performing the task execution. Two frameworks currently compose the StarSs programming model family: COMP Superscalar (COMPSs) [53], which provides the programming model and runtime implementation for distributed platforms such as Clusters, Grids and Clouds, and OpenMP Superscalar (OmpSs) [330, 136], which provides the programming model and runtime implementation for shared memory environments such as multicore architectures and accelerators (such as GPUs and FPGAs). Apart from the target platform and granularity of the tasks, they also differ in the programming languages and data dependencies supported as depicted in Figure 4.1.

Our proposal to simplify the development of parallel applications for distributed and heterogeneous systems consists of a combination of the StarSs programming models in a hierarchical way. So, an application is represented as a workflow of coarse-grain tasks. Each of these coarse-grain tasks can implement as well a workflow of finer-grain tasks. In other words, the coarse-grain

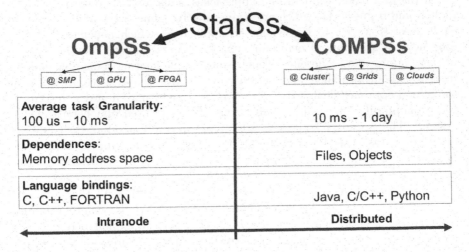

FIGURE 4.1: StarSs programming models overview.

tasks will be defined as COMPSs tasks and each COMPSs task will be implemented as an OmpSs application with fine-grain tasks parallelism.

At runtime, coarse-grain tasks will be managed by COMPSs runtime optimizing the execution on a platform level by distributing tasks in the different compute nodes according to the task requirements and the cluster heterogeneity. On the other hand, fine-grain tasks will be managed by OmpSs which will optimize the execution of tasks at node level by scheduling them in the different devices available to the assigned node.

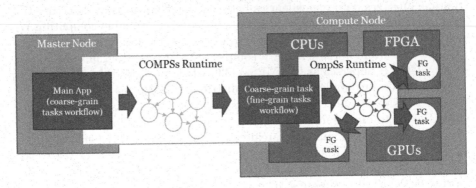

FIGURE 4.2: Application execution overview.

Figure 4.2 shows an overview of how applications are executed in the distributed environment. From the main code of the application, the COMPSs runtime detects data dependencies, builds a Direct-Acyclic Graph (DAG) with the coarse-grain tasks, and schedules dependency-free tasks according to data location and the coarse-grain constraints, deciding which task can run in parallel in each node and ensuring that the different tasks are not colliding in the use of resources. For each coarse-grain task, the OmpSs runtime (a.k.a Nanos++) builds a DAG with the fine-grain tasks and schedules them in the resources assigned by the platform-level scheduling.

To program an application with the proposed programming model, developers have to identify the parts of the code that are candidates to be coarse-grain tasks. These are usually functions repeated several times in the code and with enough computation to compensate for the overhead of spawning a remote process (around 10 ms). To indicate that a method is a coarse-grain task following the COMPSs syntax. The main code of the application can be implemented as a normal sequential C/C++ code. The similar procedure is done for fine-grain tasks. In this case, tasks can have finer granularity due to the shared memory environment. To define a fine-grain task, developers have to indicate the code which is task and the input and output data by using the OmpSs syntax. In addition, annotations for tasks that are accelerator kernels should also include a *target* clause in the directive to differentiate them from regular CPU tasks.

In addition to the normal task definition, the programming model provides mechanisms to support different tasks' versions and allocation of resources based on tasks' constraints in order to make application codes adaptable to the underlying infrastructure. Adding an *implements* annotation, developers can indicate that a task implements the same functionality of another task. So, the runtime can execute any implementation of each task according to the available resources. For instance, a part of one application can implement a feature programmed for running in a CPU, but also, a CUDA kernel implementing the same feature could be also available to run in a GPU. In this case, developers can define the CUDA kernel as a task which implements the CPU task.

In addition to different implementations, developers can also add *constraints* to tasks to specify the minimal resource requirements required to run this task type, a certain software, etc. The runtime takes these constraints into account when scheduling these tasks in resources which must fulfill these constraints. For instance, for a coarse-grain task which is composed of a set of fine-grain tasks targeting accelerators we have to add a task constraint to indicate that it requires a GPU, FPGA, etc. Or in the case of coarse-grain tasks implementing a workflow of fine-grain tasks, the constraints directive can be also used to indicate the number of cores required to run the fine-grain tasks in parallel.

The following paragraphs provide details about how COMPSs and OmpSs manage the execution in the distributed and heterogeneous environments. As use case we use a matrix multiplication, where two big matrices are multiplied by blocks because they are not able to fit in the node memory so the main block-by-block multiplication is developed as a workflow of corse-grain tasks using COMPSs, and each block multiplication is parallelized inside a node as a workflow of fine-grain tasks with OmpSs.

4.2.1   COMPSs

COMPSs is the StarSs programming model instance which aims to provide a parallel programming environment for distributed computing platforms. COMPSs tasks are annotated using an interface file, where the tasks are identified by declaring the method which implements the functionality and the directionality of its parameters. An the main workflow of tasks is implemented as a normal sequential code.

Listing 1 and Listing 2 show the COMPSs interface file and the main workflow implementation, respectively, for the blocked matrix multiplication example. Method *matrix_multiply* performs a fixed-size block matrix multiplication of $BS \times BS$ size. In the interface file (Listing 1), method *matrix_multiply* is identified as a task. In addition, the annotation indicates that this task has two input parameters (parameters that are only read—in in the annotation) and an input/output parameter (a parameter that is read and written—inout). Another task declaration example is found in the method *load_block*. In this case, it has two input parameters: size and file parameter, and a return parameter (block matrix) which has output direction by default.

Listing 1—COMPSs Matrix Multiply task definition at the interface file

```
 1  interface Matmul
 2  {
 3      float[BS][BS] load_block (in int size, in file block_file);
 4
 5      @Constraints(processors={@Processor(type=CPU, computingUnits=4),
 6                              @Processor(type=FPGA, computingUnits=1)})
 7      void matrix_multiply(in float[BS][BS] blockA, in float[BS][BS] blockB,
 8                              inout float[BS][BS] blockC);
 9
10      @Constraints(processors={@Processor(type=GPU, computingUnits=1)},
11                              software=CUDA)
12      @Implement(matrix_multiply)
13      void matrix_multiply_cuda(in float[BS][BS] blockA, in float[BS][BS]
14                              blockB, inout float[BS][BS] blockC);
15
16      @Constraints(processors={@Processor(type=GPU, computingUnits=1)},
17                              software=opencl)
18      @Implement(matrix_multiply)
19      void matrix_multiply_opencl(in float[BS][BS] blockA, in float[BS][BS]
20                              blockB, inout float[BS][BS] blockC);
21  };
```

With COMPSs, there is the possibility of setting constraints on the resources required for executing the tasks using the @constraints annotation. For instance, the method *matrix_multiply*, is specific for general-purpose CPUs,

Listing 2—COMPSs version of Matrix Multiply main code

```
1  void main(){
2    ...
3    compss_on();
4    size = BS * BS;
5    for (i=0; i<num_blocks; i++)
6      for (j=0; j<num_blocks; j++)
7        AA[i][j] = load_block (size, blockA_filename[i][j]);
8        BB[i][j] = load_block (size, blockB_filename[i][j]);
9
10   for (i=0; i<num_blocks; i++)
11     for (j=0; j<num_blocks; j++)
12       for (k=0; k<num_blocks; k++)
13         matrix_multiply(AA[i][k], BB[k][j], CC[i][j]);
14   compss_off();
15   ...
16 }
```

and requires four cores while the *matrix_multiply_cuda* method is specific for GPUs, requiring just one. We can also add some software constraint; for instance, the *matrix_multiply_cuda* task requires the CUDA runtime while the *matrix_multiply_opencl* task requires the OpenCL runtime.

An additional annotation (*@Implements*) supports the definition of multiple versions of a behaviour (polymorphism). In our case example, the methods *matrix_multiply*, *matrix_multiply_cuda*, and *matrix_multiply_opencl* perform the same operation on its parameters but using different types of resources (CPU versus GPU). As can be observed, these annotations are very useful for programming parallel applications to be executed in heterogeneous-computing platforms.

Applications described following this programming model can be executed on top of the COMPSs runtime system (see Figure 4.3). COMPSs runtime is responsible for detecting the tasks in the application, for building the task dependency graph that contains the inherent parallelism of the application, for all the data management in a distributed computing platform, and for the task scheduling and execution.

A COMPSs application is executed in a master-worker fashion. At building time the task invocations are substituted by calling to the COMPSs runtime and at the worker, a stub code is generated to decode the messages sent by the runtime and executes the task in the assigned resources. This part is compiled with Mercurium (OmpSs compiler), which detects the defined OmpSs tasks inside the COMPSs tasks binding the COMPSs worker stubs as a normal OmpSs application. The compilation is configured with the necessary Mercurium flags according to node properties and defined tasks. For instance if a node has GPUs with CUDA, the worker part is compiled with the—*ompss* and—*cuda* flags.

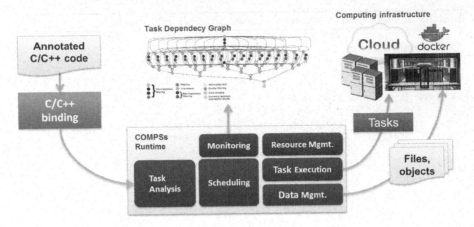

FIGURE 4.3: COMPSs runtime overview.

For each COMPSs task invocation, the COMPSs runtime analyses the dependencies with previous invoked tasks and creates a node in the task-dependency graph adding the edges with other tasks according to the detected data dependencies. Once a task is free of dependencies, the COMPSs runtime selects the fastest implementation which can be executed in the available resources according to constraints defined in the tasks' interface file. This is done by matching the task constraints with the resource properties.

Finally, if the application is running in a dynamic distributed platform such as Clouds, the COMPSs runtime is also able to self-adapt the number of worker nodes to the available parallelism in the different parts of the application [257]. In regions where there are more dependency-free tasks than available resources to execute them, the COMPSs runtime contacts the Cloud provider to request more computing nodes. In contrast, when there are idle resource, it contacts the provider to switch off this resource.

4.2.2   OmpSs

The OmpSs programming model allows the expression of parallelism that will be executed in the available resources among the host SMP cores, or integrated/discrete GPUs and/or FPGAs, in a compute node. OmpSs is very similar to OpenMP tasking, being used as a forerunner prototyping environment for future OpenMP features. On GPUs, both CUDA and OpenCL kernels are supported. For FPGAs, OmpSs uses the vendor IP-generation tools (Xilinx Vivado and Vivado HLS [281, 384], or Altera Quartus [209]), to generate the hardware configuration from high-level code. On FPGA platforms that support OpenCL, OpenCL kernels are also supported. In addition, OmpSs can also leverage existing IP cores, provided they adhere to the same interface with our software platform.

Figure 4.4 shows an example of an OmpSs application for a heterogeneous system with SMP cores, GPUs, and a FPGA. It shows a main code executing a blocked matrix multiplication by calling a *matrix_multiply* function for each block multiplication. Function *matrix_multiply*, the same method used above in *COMPSs*, is defined as a task at the compute-node level, with input dependencies *a* and *b* and input/output dependency *c*. Each call to this function will generate a task that will be run when its dependencies are ready. This task has also been defined to be potentially executed in two target devices (*target device(smp,fpga)* directive): any of the cores of the *smp* running the application and three instances of an accelerator that will be built to do this task in the FPGA. The accelerator has been tuned by the programmer to exploit the parallelism of the FPGA by using some additional directives (*#pragma HLS*) not related to OmpSs programming model. A better optimized option of the SMP code could also be used through the *implements* clause.

```
#pragma omp target device(cuda) implements(matrix_multiply)
#pragma omp task in([BS]a,[BS]b) inout([BS]c)
__global__ void matrix_multiply_cuda(float a[BS][BS], float b[BS][BS],float c[BS][BS]);

#pragma omp target device(opencl) implements(matrix_multiply)
#pragma omp task in([BS]a,[BS]b) inout([BS]c)
__kernel void matrix_multiply_opencl(float a[BS][BS], float b[BS][BS],float c[BS][BS]);

#pragma omp target device(fpga,smp) copy_deps num_instances(3)
#pragma omp task in([BS]a,[BS]b) inout([BS]c)
void matrix_multiply(float a[BS][BS], float b[BS][BS], float c[BS][BS]) {
#pragma HLS inline
#pragma HLS array_partition variable=a block factor=BS/2 dim=2
#pragma HLS array_partition variable=b block factor=BS/2 dim=1
  for (int ia = 0; ia < BS; ++ia)
    for (int ib = 0; ib < BS; ++ib) {
#pragma HLS PIPELINE II=1
      float sum = 0;
      for (int id = 0; id < BS; ++id)
        sum += a[ia][id] * b[id][ib];
      c[ia][ib] += sum;
    }
}


...
for (i=0; i<num_blocks; i++)
  for (j=0; j<num_blocks; j++)
    for (k=0; k<num_blocks; k++)
      matrix_multiply(AA[i][k], BB[k][j], CC[i][j]);
#pragma omp taskwait
...
```

FIGURE 4.4: OmpSs version of Matrix Multiply code.

Figure 4.4 also shows two other versions of task *matrix_multiply*, specified using the *implements* clause: *matrix_multiply_cuda* and *matrix_multiply_opencl*. The first one is a CUDA kernel and should be compiled offline, meanwhile the OpenCL kernel can be compiled at runtime. For FPGA tasks, those are automatically compiled offline, and OmpSs generates the programmer defined accelerators. This is done by the source-to-source compiler

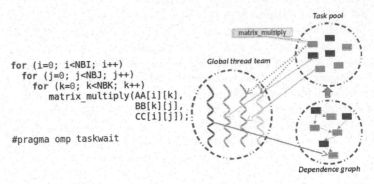

```
for (i=0; i<NBI; i++)
  for (j=0; j<NBJ; j++)
    for (k=0; k<NBK; k++)
      matrix_multiply(AA[i][k],
                      BB[k][j],
                      CC[i][j]);

#pragma omp taskwait
```

FIGURE 4.5: High-level representation of the Nanos++ environment.

Mercurium, avoiding user errors and speeding up the process of hardware generation for the supported platforms. Programmers can deactivate the accelerators' generation, to reduce the overall compilation time, if they have been previously generated [72].

On the other hand, Nanos++, the OmpSs runtime, takes care of executing tasks annotated by the programmer in the available resources. The high-level view of the execution environment is presented in Figure 4.5.

Nanos++ environment has a *thread team* created by default, the *dependence graph* used to organize tasks that still have pending data dependencies to be resolved, and the *task pool* representing all the ready tasks. Running threads create tasks and insert them into the dependence graph. When data dependencies have been fulfilled, the thread detecting this situation moves the dependence-free tasks to the task pool. When a thread finishes the execution of a task, it becomes idle and it looks for work in the task pool. The Nanos++ runtime will also take care of the possible heterogeneity expressed by the programmer and the necessary memory transfers (copies).

In heterogeneous environments, Nanos++ has a specific subset of threads that represent each of the heterogeneous devices. We call these threads *helper threads*. Figure 4.5 shows, on the left side, the code invoking the heterogeneous task *matrix_multiply* and, on the right side, the overview of threads and task pool in the runtime. The orange thread (thread number 4, on the right hand side of the Global thread team) in the figure is one of those helper threads. In this particular example, it may represent one FPGA accelerator.

Tasks executing in devices, with their own local memory, may need copy data from/to the device. In particular, the device memory space is main memory in the SMP, accessible from the accelerators and the SMP cores, physically contiguous, pinned and non-cacheable. With *copy_deps* clause the programmer indicates that all the dependences will require, at runtime, device memory space for copies between host memory and accelerators local memory. Alternatively, *copy_in/out/inout(list-of-variables-with-size)* clauses indicate the list of parameters of the task that needs to be copied to/from the accelerator and deactivate the, by default, *copy_deps* clause. In both cases, the runtime takes care of allocating device memory and copies between user and device memory for the list of parameters labeled as copies. If any of the task parameters is a scalar, it is directly passed to the accelerator. In case that copies are not requested, the programmer kernel code can access the device memory without any change. This makes programming easy and useful to apply blocking techniques from inside the accelerator.

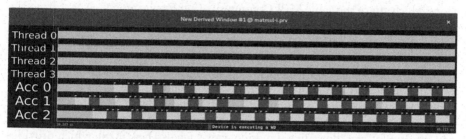

FIGURE 4.6: FPGA execution trace with three 128×128 Matrix Multiply accelerators.

OmpSs allows tracing of the execution of Nanos++ threads (running in cores) and accelerators. For the threads, it provides information at application and runtime levels so that the programmer can analyze both the application and runtime internals such as the creation of tasks, task executions, taskwaits, etc. For the case of FPGA tasks running in accelerators, current support provides the user with information of execution time of the data movements done and computational time of the kernel. Figure 4.6 shows part of an execution trace where this information is shown for three accelerators of the Matrix Multiply application. Each accelerator is running several FPGA tasks with three different stages: copy in data, compute Matrix Multiply, and copy out of the data. Those stages are shown with three different colors (light green, brown and yellow respectively) in the execution trace for each of the accelerators and task instance. The execution trace generated is done using an internal tracing library and can be visualized with Paraver [56].

4.3 Related Work

In the area of distributed computing, developers can find several high-level solutions. We can find tailored solutions for a certain type of applications which use DSLs to abstract most parallelism and execution complexity. The complicated part when programming applications with these frameworks is mapping your application to the concepts of the new model, which is not always easy. This is the case of data-analytic frameworks such as Map Reduce [127] Spark [393], graph-processing frameworks such as Pregel [265], or

Deep learning frameworks such as Tensor-flow [35]. Another widely-used option to create distributed applications is using MPI [182]. MPI applications are implemented in a Single Program Multiple Data (SPMD) fashion where the different MPI processes are executing the same code an interchange messages to transfer data between processes. Some extensions have been recently introduced to the MPI specification to support Multiple Program Multiple Data (MPMD). With these extensions developers can implement different applications which share the same MPI Communication Environment. Apart from the aforementioned options which are focusing on certain pattern, there are also other options to program general purpose applications. One of these options are workflow enactors, these tools provide an interface to describe and execute a workflow (Graphical o by means of a Workflow Description Language). This workflow description mainly combines execution of binaries in a static way. Examples of these tools are Galaxy [174], Taverna [292], Kepler [45] or Fireworks fireworks. Finally, we have also find task-based frameworks where users can implement workflows in a dynamic way, such as Dask [328], which provides a Python API to describe task-based workflow as programs, and Swift [381], which provides their own scripting language.

Most of the distributed computing approaches do not provide access to heterogeneity. Only Tensor-flow or simulators like GROMACS have version which execute the code in a GPU. So, to implement an application which supports the execution of application in distributed systems as well as using the accelerators available in the computing nodes, we need to combine two programming models. A very extended approach is the combination of MPI with OpenCl or CUDA. But in this case, developers have to implement all the OpenCL and CUDA API inside the MPI code to manage the data management and execution in the accelerators. On the other hand, in order to reduce the programmer effort, models have to provide the means to perform/indicate data transfers between host and accelerators in an easy way, allowing to reduce the impact of those communications. Our proposal addresses these challenges achieving high productivity by providing higher-level abstractions that could help the programmer generate high-performance code on them.

In the area of heterogeneous computing, the Vineyard project [374] aims at facilitating heterogeneous programming, based on OpenSPL [270], OpenCL [294] and SDSoC [385]. The Ecoscale project [142] targets applications written in MPI and OpenCL, to synthesize the OpenCL kernels for the FPGAs, and support distributed and heterogeneous computing. For both projects, the goal is to efficiently execute functionality in the FPGAs. In addition to this, we also provide heterogeneous execution on both the FPGAs and the available host cores.

PGI [315] and HMPP [134] programming models are two other approaches quite related to OmpSs. PGI uses compiler technology to offload the execution of loops to the accelerators. HMPP also annotates functions as tasks to be offloaded to the accelerators. We think that OmpSs has higher potential in that it shifts part of the intelligence that HMPP and PGI delegate in the compiler to the OmpSs runtime system. Although these alternatives do support a fair amount of asynchronous computations expressed as futures or continuations, the level of lookahead they support is limited in practice.

There are a large number of frameworks targeting the High-Level Synthesis from C/C++. Vivado HLS [384] is the Xilinx tool that we use from OmpSs to generate the FPGA IP blocks. Xilinx SDSoC [385] runs on top of Vivado HLS, and better integrates the execution environment for the Xilinx Zynq platform (7000 and Ultrascale+), with the automatic generation of the complete Linux system for the target platform. LegUp [88, 157] synthetizes C code with Pthreads and limited OpenMP annotations. Each thread (code) is synthesized as an accelerator at compile time. The remaining (sequential) portions are executed in the processor, invoke accelerators and use synchronization functions to retrieve their return values. In OmpSs, the accelerators are also generated at compile time but they correspond to tasks with target device FPGA. Indeed, it is also possible to specify tasks that run in the SMP, that can run in both SMP or FPGA, or that can substitute other tasks. OmpSs runtime takes care of issuing data transfers and task executions among the cores and the IP accelerators based on the readiness of the task dependencies, if defined.

4.4 Conclusion

This chapter presents StarSs, a family of task-based programming models where developers define some parts of the application as tasks indicating the data required by those tasks and their direction. Currently, two frameworks compose the StarSs programming model family: COMP Superscalar (COMPSs), which provides the programming model and runtime implementation for distributed platforms such as Clusters, Grids, and Clouds, and OpenMP Superscalar (OmpSs). StartSs approach affords and simplifies the parallel programming of distributed heterogeneous systems, presenting different advantages with respect to other approaches:

First, it provides a programming model which integrates distributed computing with heterogeneous systems, allowing developers to implement parallel applications in distributed heterogeneous resources without changing the programming model paradigm. The programmer does not require to learn different programming models and APIs. Programmer only needs to annotate which parts of the program are tasks, if they use input or output task data and the task granularity level.

Second, developers do not have to deal with data transfers, like in MPI. The programming model runtime analyzes data dependencies at distributed and node levels and keeps track of the data locations during the execution.

So, it tries to schedule tasks close to the consumed data and when this is not possible, it transparently transfers the required data.

Third, we have extended the versioning and constraints capabilities of these programming models to make adaptable applications for distributed and heterogeneous computing environments. With these extensions, developers are able to define different versions of tasks for different computing devices (CPU, GPUs, FPGA) or combinations of them. So, the same application will be able to adapt its execution to the different resource capabilities of the heterogeneous platforms without having to modify the application.

Fourth, our system facilitates the FPGA programmability and speeds up all the process of software/hardware design space exploration on FPGA-based systems, avoiding hardware design errors. In particular, at compile time, FPGA accelerators are automatically generated based on programmer annotations, using vendor HLS tools. At runtime, the runtime library takes care of all the synchronization and management of the FPGA accelerators, transparently to the programmer.

To the best of our knowledge, the combination of COMPSs and OmpS is the first attempt of this kind of dynamic work distribution across compute nodes, SMP cores, and other devices as GPUs or FPGAs that integrates previous programmability features.

5

Design-Time Tooling to Guide Programming for Embedded Heterogeneous Hardware Platforms

R. De Landtsheer, J. C. Deprez, and L. Guedria

Centre d'Excellence en Technologies de l'Information et de la Communication (CETIC), Gosselies, Belgium

CONTENTS

Heterogeneous hardware platforms are now available in many devices. However, they remain extremely complex to program, and it is even harder to exploit the power of heterogeneous hardware platforms adequately. Tools for assisting analysts-developers at design time to make efficient design space exploration for better exploiting heterogeneous hardware architecture are therefore truly needed. This work presents Placer, a model-based tool for optimizing the mapping, i.e., placement and scheduling, of task-based software onto heterogeneous hardware, and DS-Explorer for rapidly prototyping software tasks offloading on FPGA and obtaining static and dynamic properties of software tasks execution on FPGA. Placer and DS-Explorer are not implemented from scratch. They, respectively, leverage on existing tools Quickplay from Accelize and OscaR by UCL and CETIC. An application of both tools, Placer and DS-Explorer is demonstrated in an industry-use case study named Aquascan provided by Deltatec.

5.1 Introduction

Designing and developing optimal software architectures for embedded applications to execute on a set of heterogeneous hardware devices to achieve best time and energy performance requires making complex decisions [79]. Decisions on placing software components on the appropriate hardware processing devices must not just consider processing performance but also data transfer. On multi-core and many-core devices, there is not much constraint on the code-writing exercise. On the other hand, when GPU and FPGA are available in an embedded device then specific code must be developed for these types of hardware. Determining what software components would benefit most from executing on GPU or FPGA is extremely hard. Mainly because it is not always the intuitively heavily parallelisable software task of an application that should automatically be placed on GPU or FPGA to achieve the desired time and energy performance. Importantly, the type and size of data transfer must be considered to make an optimal placement decision. Otherwise what may seem like a judicious placement of a heavily parallelisable task on a GPU or a FPGA could yield suboptimal execution results, for instance, if datasets to transfer from the main platform to a GPU or FPGA are large or if datasets do not generate a constant predictable datastream. Effort for automatically-generating efficient GPU execution from standard C code starts providing encouraging results [183]. On the other hand, design-space exploration for FPGA has so far remained mostly a manual effort heavily influenced by the designer's experience even if graphical tools such as PREESM [318] [308], ORCC [390], or QuickPlay by Accelize [20] may help with the exploration exercise.

Furthermore, next to time performance, in embedded systems composed of several devices where some run on battery while others are powered through a UPC, energy consumption must often be considered at least for battery-operated devices. Thus, even expert developers cannot reliably answer whether a given software-component placement and software-task schedule provide the most appropriate trade-off between time and energy performance. To ease this design-space exploration effort, the H2020 European TANGO project has developed design-time tools [356].

A first tool, DS-Explorer facilitates several aspects related to the exploitation of hardware platform with FPGA, notably actions for automating the exploration of certain design variants when compiling a software task on a FPGA (e.g., cloning parallelisable subtasks on the FPGA), and second, for capturing time and energy characterisation of the different software task variants when executed on a FPGA.

A second tool, Placer, optimizes the mapping (placement and execution scheduling) of software tasks onto heterogeneous hardware [325][241].

Placer inputs a model of hardware processing, storage, and transfer capabilities alongside a model of computation and communication (MoCC) of software tasks and their data exchanges. The MoCC provided by Placer makes it possible for a software task to provide several implementations, for instance, for executing on different hardware-processing element such as CPU, GPU, or FPGA. Furthermore, the MoCC is augmented with different software-task dependencies information, as well as other properties, such as time and energy performance of each task implementation on a given hardware. From these input models, Placer computes the optimal placement and schedule of software task on the given hardware elements.

This chapter presents the usage of tools throughout the design exploration process, notably Placer that can be used several times at different moment of the design phase. After the investigation of a few what-if scenarios for exploring what tasks to port on FPGA would benefit most the overall application execution behaviour, DS-Explorer will then be used for the obtaining first rapid prototypes to execute on FPGA.

In other words, the contribution of this work is to enable a breadth-first design-space exploration through easing static and dynamic characterisation of software tasks variants on FPGA-augmented platforms together with the power of Placer for exploring the optimal placement and scheduling of these tasks to achieve the desired trade-off between time and energy performance.

The second contribution presents the usage of Placer and DS-Explorer on an Industry Use Case involving image analysis and optical digit recognition on a small embedded system with heterogeneous hardware.

This chapter is organised as follows. Section 5.2 presents Placer's input and output metamodels. Section 5.3 illustrates briefly how these two metamodels have been implemented into Eclipse Modelling plugins. Section 5.4 describes the Placement algorithm used by Placer and based on a Constraint Programming approach. Section 5.5 describes the DS-Explorer tool. Section 5.6 present the Industry-use case and shows that Placer and DS-Explorer can be used to assist analysts/developers in their design-space exploration for exploiting the power of an underlying small heterogenous-hardware platform.

Section 5.7 presents the related works. Section 5.8 provide the conclusion and future directions for evolving Placer and DS-Explorer.

5.2   Placer—Input and Output Metamodels

This section describes the main features of the input problem supported by Placer. Here, we focus on the structure of the model exclusively. The concrete syntax read by Placer is based on a JSon format and is documented in the reference document [326]. In the concrete representation, each entity is referred to by its name, which is not represented explicitly in the next figures for simplicity.

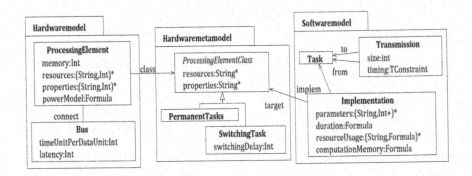

FIGURE 5.1: Input model of Placer.

The input problem sent to Placer is a model that that can be divided into four parts: the hardware metamodel defines what is an FPGA, what is a CPU, etc., and makes it possible to define new types of such processing element classes; the software model defines the software tasks and transmissions to be placed and scheduled; the hardware model defines the targeted hardware with its buses and processing elements, and the constraints and objectives defines additional constraints, such as deadline, and specifies what is to be minimized. The hardware meta-model, hardware, and software models are shown in Figure 5.1.

The output of Placer is a mapping. For each task it specifies what implementation is to be used, on which processing element it is placed and when it is to be executed. Similarly, for each transmission, it specifies on which bus it is placed an when it can take place.

5.2.1 Hardware Meta-model

The hardware meta-model defines the classes of processing elements to consider, with their specific attributes. For instance, one defines what a FPGA is at this layer, and what are the resources to consider (gates, multipliers, etc.). There are two possible classes of processing elements, namely PermanentTasks that permanently hosts tasks, such as FPGA, and runs them on demand, possibly in parallel and SwitchingTasks that execute different tasks in sequence, one at a time. For both, we can define resources that are allocated to tasks when they are hosted on the processing element, and properties that can be defined for each processing element instance. An example of property would be the operating frequency of a CPU.

5.2.2 Software Model

The software model defines tasks, their *implementation and transmissions* between these tasks. Tasks can have one or more implementations defined by their target, parameters, durations, computation memory, and resource usages. The target of an implementation must be one of the processing element classes defined at the meta-level. Implementations can be parametric; they can have set of parameters, and these parameters can be referenced in their duration, computation memory and resource usage. A parameter is defined by its name, and the set of possible value for the parameter. These parameters must be set by Placer, as part of the optimization process. Duration, computation memory, and resource usage are defined as formulas. These formulas can refer to parameters of the implementations as well as properties of the target processing element class. The resource usage declares, for each resource of the processing element class, the amount of this resource that the implementation will use during its execution for the execution in the case of SwitchingTask target or throughout the whole schedule in the case of PermanentTasks target. The computationMemory is the memory used by the implementation. The memory of processing element is considered as a particular resource because it is used both as computation memory and as buffer memory for transmission.

Transmissions are about transmitting data between tasks. A transmission can start as soon as the emitting task completes and must complete before the receiving task can start. Transmission have a size attribute, specifying the amount of data to be transmitted. Transmissions can be constrained about when they take place. There are four possible time constraints (TConstraint): ASAP: the transmission must start as soon as the emitting task has completed; ALAP: the receiving task must start as soon as the transmission completes; Sticky: Placer chooses between ASAP and ALAP; Free: meaning that they can start any time after the end of the transmitting task and must finish before the start of the receiving task.

5.2.3   Hardware Model

The hardware model defines the processing elements in the hardware and the busses that connects them. The processing elements must be of a declared processing element class and must specify values for the resources and properties declared in the processing element class. Memory is also declared at the level of the processing element and Placer ensures that the needed amount of data is available for hosting the computation memory of executing implementations and for buffering data before and after transmissions take place. There are a few variants of buses supported by Placer, in this paper, we only consider the classical half duplex buses where every connected processing element can transmit to every other connected processing element, but only one transmission can happen simultaneously.

5.2.4   Constraints and Objective

Additional constraints can be specified on this mapping problem including:

- **powerCap** specifies a maximal value for the power used by the hardware throughout the schedule. This power is the sum of the power consumed by each processing element.

- **maxMakespan** specifies a moment in time such that all tasks must end before this moment in time—energyCap specifies a maximal amount of energy that the placement/schedule requires to be executed. runOn and notRunOn specify that some tasks must (resp. mustn't)—run on some processing element.

- **samePE and notSamePE** specify that some set of tasks can be constrained to be placed on the same (resp. different) processing element.

- **mustBeUsed and mustNotBeUsed** specify that some processing element must (resp. must not) be used by at least one (resp. any) task.

    Several of these can be specified at the same time (although several power-Cap would result in posting only the tightest powerCap). Some combination of constraint can make the mapping infeasible. End user can somehow interact with Placer by running it several times with different constraints on each run.

    The objective functions supported by Placer include:

- **sat** to just ask Placer to find a feasible mapping.

- **minMakespan** specifies that the placement should minimize the makespan, that is the end time of the last task to be executed.

- **minEnergy** specifies that the placement should minimize the total energy consumed for executing it.

- **minEnergyAndMakespan** specifies that Placer should search for the Pareto-optimal set of placement that minimize both makespan and energy.

5.2.5   Output Model of Placer

The decision variables defined in the model and for which Placer needs to find a proper value constitute a mapping, and they can be divided into three sets:

- **Selection** of the implementation for each task. In case the selected implementation has parameters, these must also be assigned a value selected in their defined range.

- **Placement** of tasks on the processing element and routing of transmissions on the hardware buses.

- **Scheduling** of transmission and task executions. The schedule is made of the start and end time of each task and transmission.

5.3   Graphical Front-End

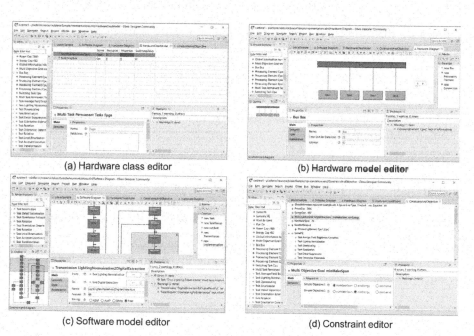

(a) Hardware class editor

(b) Hardware model editor

(c) Software model editor

(d) Constraint editor

FIGURE 5.2: Graphical editor of Placer.

To make Placer more usable, a graphical front-end is developed, based on Sirius technology [357]. Such technology makes it possible to quickly develop graphical editors for "boxes-and-arrows"-based models from a given meta-model, like Placer's meta-model in Figure 5.1. The editor itself is an Eclipse plug-in and benefits from all the support of the Eclipse framework in terms of view, windows, pop-up menu, etc. [141]. Screenshots of our editor are given in Figure 5.2. They show the three views corresponding to the software, hardware and hardware class, and it also shows a view where the user can specify constraints and objective according to the vocabulary supported by Placer.

There are a few tools offering graphical support for handling similar models [30][318]. We developed our own graphical tool because all solvers and graphical editor tools have a different modelling language, and the best way to expose all the features supported by Placer, and only these features is to propose a tailored graphical front-end. Besides, the cost of developing such front-end is rather negligible given the highly productive support of the Sirius framework that we had selected. Concretely, the front-end was developed as the result of an M.Sc internship of 14 weeks covering the learning of Placer, Sirius, and the development of the front-end itself.

The editor supports a few views of the global input model, namely:

- Hardware meta-model that lists the processing element classes.

- Software diagram that displays tasks and transmissions; these concepts can be edited through this diagram.

- Hardware diagram that displays the processing elements and transmissions; these concepts can be edited through this diagram.

- Constraints and objectives that lists properties of files (data unit and time unit), additional constraints.

- The objective function.

- Besides it also comes with the support for undo-redo, saving, and functionalities have been developed to validate the structure of the model, such as checking that all tasks have different names, task graphs have no loops, etc.

- Some concepts and attribute are accessible by means of a property window displayed below each editor and follow the focus of the mouse, as usual in such editors.

5.4 Placement Algorithm

The optimization engine is built on the OscaR.cp constraint programming engine [299]. Constraint programming (CP) is a declarative framework for combinatorial optimization where the optimization problem can be specified using variables, in this case of integer type, and constraints. Placer formulates the mapping problem as a constraint programming (CP) problem and uses the CP engine to solve this constraint problem. The solutions of the CP problem are then translated back into mappings.

5.4.1 Expressing the Mapping as a CP Problem

A constraint programming engine offers two main classes of primitives, namely: decision variables, which are the variable that we look a value for, and constraints. In our case, there are only variables of integer type, with a range of possible values that must be provided to the engine and must be as small as possible. A library of constraint is provided with the constraint programming engine, and covers basic mathematical operations (+, -, >, max, etc.) as well as more sophisticated constraints, such as resource constraints used for scheduling.

Within Placer, the mapping problem is formulated as a scheduling problem, where the (primitive) tasks of the scheduling problem, include the tasks of the software and the transmissions.

Precedence constraints are posted between these primitive tasks to ensure that transmissions do not start before the emitting software task has completed, and that software tasks do not start before all their incoming transmissions have arrived.

Software tasks have four additional decision variables, namely the processing element where they are located, the implementation they have, their duration, and the power they consume throughout their execution. These four variables are linked together through a table constraint that restricts these four variables to the set of feasible quadruplets. This set is computed from the available implementations, their target, and the available processing elements.

Transmission tasks have two additional decision variables, namely the duration of the transmission and the bus where the transmission is routed. Again, these two variables are constrained to belong to a set of couples, computed from the available hardware buses, and their speed characteristics combined with the size of the transmissions. The bus is further constrained to be adjacent to the processing elements of the emitting and receiving tasks through a table constraint on these three placement variables. These triplets are extracted from the adjacency properties of the buses from the hardware model.

Some of the table constraints mentioned here were split into smaller constraints because they combine variables with very large range, such as duration that can range from zero (for transmission within the same processing element) to very large values, as in the case study of this paper. Table constraints do not perform well on variable with a large range of possible values, so that all timing and energy variables had an additional ID variable. This ID was incorporated into a table constraint combining ID, processing element, and implementation. Additional constraints are used to bind the value of these ID variables to the actual timing and power variables. We used the element constraint to do this: it inputs an array of constant values (such as the duration of a task for each of its implementation for each processing element of the considered hardware), and an ID and output variable, and constraints the output variable to be equal to the value in the array denoted by the ID variable. This does not impact the semantics.

CP engines reason on variables by representing somehow the set of values they can be assigned to each of them and try to shrink these sets by exploiting the constraints. CP engines, including OscaR.cp, support two possible representations for these sets of possible values, and the user must select the appropriate one every time. The two possible representations are: the actual set of possible values or, with some loss of precision, an interval [min;max]. The representation by set is used in Placer for all placement variables (target, implementation selection, transmission routing), and interval representation is used for all variables dealing with time or energy because their range of possible values is so large that it is not practical to represent all the possible values in a set.

Switching-tasks processing element (resp. buses) represent standard scheduling resources with fixed instruction sets such as CPU or GPU. Any software tasks (resp. transmissions) can be scheduled on them during software execution.

Multi-tasks processing elements represent resources that can only be configured once prior to sofware application execution, as it is usual for FPGA. Only certain software tasks with an appropriate implementation can schduled on multi-tasks processing elements. Their model is a multi-dimensional knapsack. The dimensions of the knapsack are the resources of the processing element, and the resource consumption of the implementations. The selected items are the ones that are placed on the processing element.

5.4.2   Solving the CP Problem

CP is an exhaustive symbolic approach. It will try all and every combination for the values of the decision variables, but lots of these combinations are bundled into sets that are represented using some interval on the possible values of each variable and evaluated at once. Furthermore, CP can reason on these sets of variables to exclude non-interesting solutions (infeasible or not optimal) by shrinking the interval of possible value for each variable. This is what makes CP a much more efficient technique than brute-force enumeration.

Beside the model, a search heuristic must be specified to guide the exploration of the search space. This is the tricky part of CP engines: search strategies are hard to elaborate yet play a key role in the overall efficiency of the search engine.

These strategies are intertwined with the propagation mechanism, so that setting the value of some variable through the search strategies might cascade to other variables. For instance, placing a task on a processing element will likely set its implementation and its duration; it will also constrain the outgoing transmission to take place on some bus, depending on the placement of the receiving task; incoming transmissions are constrained as well, etc. In Placer, a set of primitive search strategies are available, and a global search strategy is specified as a sequence of these atomic search strategies.

A strategy is generally about setting the value of a variable. All strategies include two dimensions: one that specifies in which order the variables are assigned, and the other that selects the order in which their possible value is to be tried. The available primitive search strategies are as follows:

- **Transmission Routing**: decides for each transmission on which bus it is to be routed; the largest transmission is placed first, and the bus where it is faster is selected first. Notice that local transmissions where the source and target processing element require zero time unit, so that this heuristic will tend to use local transmissions first, reverting to transmissions on buses in a second attempt

- **Task And Transmission Starts**: this is a pure scheduling heuristics that has been developed by the CP community. It sets the start time of each task and transmission, and sets this decision for the tasks and transmissions that take more time. It also uses some form of learning to further identify the tasks and transmissions that play a critical role in the final schedule [169].

- **Task Placement Less Buzy Proc First**: places the tasks on processing elements; selects the ones with the longest possible duration first, and puts them on the less-buzzy processing element first

- **Local Or Bus Transmission Largest First Local First**: decides for each transmission on which bus if the transmission has to be local or needs to be routed on a bus. It makes this decision on transmissions by increasing sizes, and first tries routing it as a local transmission. This heuristic does not select the final bus.

- **Local Or Bus Transmission Longest Adj First Local First**: decides for each transmission if it will take place as a local transmission, that is the source and target tasks are located on the same processing element and the transmission, does not require an actual bus, or if the transmission is taking place on a bus that connects the transmission and receiving tasks. The heuristic tries the local case first, and decides for each transmission in decreasing transmission size.

- **Task Placement Fastest Implem Plus Less Buzy Proc First**: assigns tasks to processing elements. The tasks are placed in decreasing duration, and the decision places tasks on the processing element where it is the fastest first.

Placer search a solution by applying a sequence of the strategies listed above. While a built-in sequence is provided in Placer, it is possible to overwrite it by a new specified on the command line when running Placer.

Placer can be run in a pure CP fashion. In this case, the computed results are guaranteed to be complete. This algorithm is, however, rather slow and is only applicable to small examples.

A CP engine can be used either in a sat mode or in a minimization mode. These modes are identified from the objective function specified in the input file. When used in minimization mode, it performs a branch and bound search on the objective function.

5.4.3 Large Neighbourhood Search (LNS)

LNS is a local search approach where a best-found solution so far is stored in memory, and CP search is repeatedly performed on the original problem where a fraction of the decision variables is set to be equal to the values they have in the best solution found so far.

A synthetic algorithm of LNS is shown in Algorithm 2 where findFeasible Solution and findBest represent calls to the CP solver in sat and in minimization mode, respectively, and selectSomeAssigments represents the heuristic procedure that selects some assignments from the best solution that are injected when calling the CP solver through findBest. The findBest function is generally bounded in such a way that it does not spend too much time in case the selected assignments do not make it possible to reach better solutions quickly. Typically, the number of backtracking performed by the CP engine is bounded, and when this bound is reached, the CP engine terminates without completing its exploration, so that another LNS iteration can be started.

This approach greatly improves on the runtime of Placer, at the cost of losing its formal completeness. Nevertheless, the delivered solutions are still of very high quality.

The stopCriterion is parametrized through the command line and incorporates a maximal number of iteration and a maximal number of consecutive iteration without improvement; the search stops as soon as any of these criteria is met.

The procedure selectSomeAssigments randomly selects placements with a probability of 10% for each software task, and no assignment related to scheduling. LNS is a local search approach where a best-found solution so far is stored in memory, and CP search is repeatedly performed on the original problem where a fraction of the decision variables is set to be equal to the values they have in the best solution found so far.

A synthetic algorithm of LNS is shown in Algorithm 2 where findFeasible Solution and findBest represent calls to the CP solver in sat and in minimization mode, respectively, and selectSomeAssigments represents the heuristic procedure that selects some assignments from the best solution that are injected when calling the CP solver through findBest. The findBest function is generally bounded in such a way that it does not spend too much time in case the selected assignments do not make it possible to reach better solutions quickly. Typically, the number of backtracking performed by the CP engine is bounded, and when this bound is reached, the CP engine terminates without completing its exploration, so that another LNS iteration can be started.

This approach greatly improves on the runtime of Placer, at the cost of losing its formal completeness. Nevertheless, the delivered solutions are still of very high quality.

The stopCriterion is parametrized through the command line and incorporates a maximal number of iteration and a maximal number of consecutive iterations without improvement; the search stops as soon as any of these criteria is met. The procedure selectSomeAssigments randomly selects placements with a probability of 10% for each software task, and no assignment related to scheduling.

Algorithm 2 Large Neighborhood Search

---
currentSol = findFeasibleSolution(problem)
while ! StopCriterion **do**
    partialAssignment = selectSomeAssigments(
                    currentSol)
    newSol = findBest(problem
                $\wedge$ partialAssignment
                $\wedge$ obj < currentSolution.obj
                ,obj)
    if newSol.isDefined **then**
        currentSol = newSol
    end if
end while

5.4.4 Controlling the Search

The search strategy of Placer can be controlled through the command line through the following parameters:

- The CP search strategy can be specified as a sequence of basic search strategies proposed by Placer

- Maximal runtime

- Whether pure CP or LNS must be used

- Bounds on the number of iteration

- Smart stop criterion for the LNS based on the number of consecutive LNS iteration without improvement

- Bound on the runtime or spent time in each LNS iteration

- The probability to use when chosing which variable to relax and which variable not to relax; by default, this is set to 10%

An exhaustive help documentation is available through the—help command line parameter.

5.5   DS-Explorer

DS-Explorer is intended for design space exploration at a higher level than standard tools capable of transposing C code to FPGA executable kernels such as found in the SDSoC from Xilinx [385] or Quickplay from Accelize [20]. In particular, DS-Explorer aims at easing the exploration of design space alternatives when FPGA acceleration is involved.

DS-Explorer requires an initial baseline QuickPlay project generated by the QuickPlay tool suite. Subsequently, relying on design convention annotation, DS-Explorer will generate various derived configurations for new QuickPlay projects from the initial project. DS-Explorer can then handle their compilation and execution by invoking the QuickPlay tool under the hood through its command line interface. From the compilation and execution, it is possible to obtain characterisation data for each alternative design. For instance, energy-performance estimation or size properties of each design alternative can already be obtained after the compilation step. The execution step on various representative input data set for the given application can then help to capture execution time data to help developers and designers characterise time performance of each design alternative generated by DS-Explorer.

Below a generic illustration of how DS-Explorer works is presented.

Current version of DS-Explorer is a command line tool that realizes the following steps:

1. DS-Explorer parses the configuration files of the QuickPlay project that is passed as a parameter and extracts the relevant configuration parameters that can be tuned, modified, or adapted to explore various derivatives from the original design.

2. DS-Explorer exposes those parameters to the user and asks them about the variations they would like to explore through a series of questions and answers (Q&A).

3. Based on the user answers, DS-Explorer generates clones of the original design with the new parameters as specified by the user.

4. DS-Explorer compiles, builds, and implements the different generated QuickPlay designs. This will produce various reports, of which the power, timing, and resource utilization reports are the most important.

5. DS-Explorer parses output reports and formats relevant characterization data about power, timing, and resource utilization for the different implementations of the FPGA kernel in a raw output file. These data will serve as attribute values in the JSON input file for the Placer tool. Future versions of DS-Explorer will handle automatic generation of JSON file generation.

Shown below is a quick overview of the terminal output from a DS-Explorer execution on a simple example. We launch the tool by giving the folder of a QuickPlay original design as a parameter:

```
> ./ds-explorer.py My_QP_Project_root_folder
```

DS-Explorer will first identify the available configuration files in the project:

```
Yaml files count is 14
```

It will then identify global parameters that can be tuned by the user. These parameters will be the basis of derivative generates QuickPlay designs:

```
The byte width is 32
The clock frequency is 100
The FIFO size is 230400
..
...
```

DS-Explorer then starts the Q&A session with the user to capture tuning constraints:

```
Would you like to change the byte width?(Yes/No): yes

The byte width takes the following values: 8, 16, 32, 64

Enter the new byte width: 16

Would you like to change the clock frequency?(Yes/No): yes

The clock frequency takes the following values: 50, 75, 100, 200

Enter the new clock frequency: 50

Would you like to change the FIFO size?(Yes/No): no
..
```

Once user tuning constraints have been captured, DS-Explorer continues processing, as explained above, until the output file containing the characterization data is generated:

```
Process finished with exit code 0
..
FPGA Utilization summary
```

Site Type	Used	Fixed	Available	Util\%
CLB	8481	0	30300	27.99
LUT as Logic	41916	0	242400	17.29
LUT as Memory	456	0	112800	0.40
Block RAM Tile	209	0	600	34.83
DSPs	0	0	1920	0.00
GLOBAL CLOCK BUFFERs	12	0	480	2.50
PCIE_3_1	1	1	3	33.33

...

FPGA Power summary

Total Power (W)	3.498
Dynamic (W)	2.941
Device Static (W)	0.556

From this characterization example, a user can consider that the FPGA implementation for the offloaded function would roughly require 28% of FPGA logic resources and 35% of embedded memory.

Giving the power figures generated above and with an additional information on average execution time on FPGA for the generated kernel (this feature is not yet fully supported/generated in current DS-Explorer version), DS-Explorer computes the average dynamic and static energy required to execute an FPGA design.

FPGA resource utilization and energy consumption are the properties to passed to the software model used as input by Placer presented in the previous sections. DS-Explorer output date will enable Placer to consider. For example, various design alternatives could provide the needed data to Placer to determine the optimal number of replications for a given FPGA-offloaded function.

5.6 The AquaScan Case Study

This section illustrates the capabilities of Placer on a simple heterogeneous embedded system.

5.6.1 Defining the Problem

The application involved in this use case is an image-processing application that reads the value indicated by a water meter (counter index) identified by a barcode. This requires processing the image to retrieve the value of digits displayed by the image of the counter and sending this result on some network.

The hardware platform is a Freescale evaluation board equipped with an i.MX6 processor. This processor is a four-cores ARM-based processor running at 1 GHz. In this section, we will consider the possibility of extending this board with an FPGA.

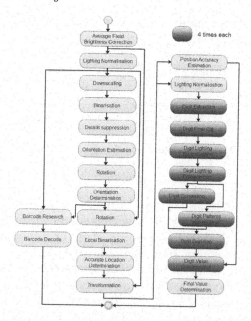

FIGURE 5.3: Aquascan task workflow.

The Aquascan software is task-based and presented in Figure 5.3. It is roughly divided into two part, the first part (on the left) is a more less linear succession of tasks and transmissions. The second part (the darker tasks on the right) is replicated four times (consider a fork before the first black task and a merge after the last black task) and can be executed in parallel. There is one instance of the right part per digit to read on the water meter, and there are four digits to read.

5.6.2 Modelling the Aquascan in Placer Format

First, we define the units to be considered globally by Placer. They are not analysed at all by Placer; they solely serve as a convention in the file, and it is the responsibility of the user to ensure that all formula or values are coherent with these units.

```
1  "timeUnit":"us",
2  "dataUnit":"8*32bit",
```

These units are selected for the duration estimates, both for the task duration and for the data transfer, to be expressed in integer dimensions without too much loss of precision. There were issues encountered with these units, as discussed at the end of this section.

The considered hardware features the i.MX6 CPU with four cores running at 1 GHz. The experiment was exclusively focusing on the four CPU cores and their communication hardware. In Placer format, we must first define the classes of processing elements that are considered. Cores are executing one task at a time, and they can switch from one task to the other task. They have no declared switching delay. We can define what a core is using the following template:

```
1  "switchingTask":{ "name":"core",
2  "resources":[],
3  "properties":[], "switchingDelay":0
4  }
```

The hardware of the experiment is made of four cores in a single chip. These are interconnected through a very fast on-chip bus. Each core has a cache, and they also have some shared caches. This shared cache cannot be modelled in Placer; however, the local cache of each core can be modelled as local memory of each core. This restriction is not so important in the Aquascan case study because the required memory for each task largely fits into the local cache of each core, so that the shared cache is virtually not used. We declare four cores, (core1 core4) in placer format. Core1 is declared as follows:

```
1  {
2  "processorClass":"core",
3  "name":"Core1",
4  "memory":32768
5  }
```

Other fields are available in Placer format; since they were not used here, they are hidden for conciseness of the report. The four declared cores are on the same chip, so that they are interconnected through very high-speed communication hardware. This hardware is represented through a single half DuplexBus that connects the four cores. The throughput of the bus is expressed in timeUnitPerDataUnit and must comply with the units defined here above.

```
1  "halfDuplexBus":{
2  "name":"globalBus",
3  "relatedProcessors":["Core1","Core2",
4  "Core3","Core4"],
5  "timeUnitPerDataUnit":1,
6  "latency":1
7  }
```

The Aquascan includes 51 tasks and 62 transmissions between these tasks. The tasks are directly instantiated from the available task graph. Each task has a single implementation targeting cores. Two additional tasks were added, namely input and output to symbolize IO operations at the extremities (entry and exit) of the overall computation performed by AquaScan. The two first tasks are as follows:

```
 1  {
 2  "name":"Input",  "implementations":[
 3  {
 4  "target":"core",
 5  "computationMemory":"2",
 6  "duration":"1"
 7  }]
 8  },{
 9  "name":"AvgFieldBrightnessCorrection",
10  "implementations":[
11  {
12  "target":"core",
13  "computationMemory":"2",
14  "duration":"934"
15  }]
16  }
```

The transmissions are declared similarly using the adequate primitives of Placer. Two transmissions are shown here:

```
 1  {
 2  "name":"InputToAvgFieldBrightnessCorr",
 3  "from":"Input",
 4  "to":"AvgFieldBrightnessCorrection",
 5  "size":8192,  "timing":"Sticky"
 6  },{
 7  "name":"AvgFieldBrightnessCorrToLighNorm",
 8  "from":"AvgFieldBrightnessCorrection",
 9  "to":"LightingNormalisation",
10  "size":8192,  "timing":"Sticky"
11  }
```

Additional constraints are meant to provide speedup to Placer as well as a means to explore some specific design on purpose. In the Aquascan, the four cores are identical, so that every permutation of the core is equally feasible. This artificially increases the size of the search space explored by Placer. To avoid this problem, we can specify that these four cores are symmetrical, and Placer adds an additional constraint to ensure that it will not waste too much time exploring symmetric mappings.

```
 1  "symmetricPE":["Core1","Core2",  "Core3","Core4"]
```

A last constraint that we can add is to force some tasks to execute on the same core. The beginning of the Aquascan software is made of a linear succession of tasks with a single implementation targeted to standard CPU cores. All these tasks can therefore be put on the same core. We can therefore specify the additional constraint as follows:

```
1  "samePE":[
2  "LightingNormalisation",
3  "Downscale",
4  "Binarisation",
5  "DetailsSuppression",
6  "OrientationEstimation",
7  "FirstRotation",
8  "OrientationDetermination"
9  ]
```

As mentioned initially, the only goal given to Placer is to minimize the makespan. This is specified as follows:

```
1  "goal":{
2  "simpleObjective":"minMakespan"
3  }
```

5.6.3   Minimizing Makespan

Once the input model is specified in JSon format, we run Placer to generate a mapping (placement + scheduling). In the following sections, we only report the placement and mention the makespan of the best schedule associated with this placement. With the above model, Placer delivers the following placement:

```
1  Mapping(makespan:82971){ Core1:82944: Input(dur:1),
2  AvgFieldBrightnessCorrection(dur:934),
3  Digit1Extraction(dur:13),
4  Digit2Extraction(dur:13),
5  Digit3Decision(dur:23),
6  Digit4Extraction(dur:13), ...
7  Output(dur:1) Core2:44571: Digit1FinalCut(dur:131),
8  Digit2LightingNormalisation(dur:33),
9  Digit3Extraction(dur:13),
10 Digit4FinalCut(dur:131), ...
11 Core3:44059:
12 Digit2PatternsIdentification(dur:854),
13 Digit3Correlation(dur:43205) Core4:46664: BarcodeResearch(dur:1425),
14 Digit1LightingNormalisation(dur:33), Digit2FinalCut(dur:131),
15 Digit3LightingNormalisation(dur:33),
16 Digit4PatternsIdentification(dur:854),
17 FinalValueDetermination(dur:1) globalBus:9575
18 }
```

The placement reports the global makespan and the placement of each task, grouped by processing element. For each processing element, it also reports the total amount of time where it was executing tasks. The durations are reported in the time unit specified in the input file; here we specified microsecond (see Section 5.2). The placement reflects some trade-off between the duration of communication, which can vary according to the amount of data to be transmitted, and the actual duration of tasks.

5.6.4 Characterising Aquascan Tasks on FPGA with DS-Explorer

Human intuition could be used to provide realistic data to Placer for properties achieved when offloading certain Aquascan software tasks on a FPGA. However, even experts usually like to adjust their intuition from actual characterisation data obtain by transposing C code into FPGA-executable kernels. For instance, even if a piece of C code has not be implemented to achieve best FPGA performance, transposing, compiling, and executing it on an actual FPGA relying on DS-Explorer and Quickplay brings an interesting baseline on time and energy performance. From this baseline, an expert designer can, in general, provide a more accurate estimate on the actual potential performance that could be achieved if further effort was spend tailing the C code to take FPGA execution in mind.

Thanks to rapid prototyping of various design alternatives, DS-Explorer provides, among others, **duration** baseline data for Placer to start from. If an expert developer believes better performance could be achieved, nothing prevents to adjust **duration** baseline data in the input model to Placer. Table 5.1 illustrates the basic characterisation feature provided by DS-Explorer. It shows that offloading software tasks `ReduceImage` or `Binarised Buffer` would currently yield execution time around 1.299 and 0.960 milliseconds and generate and average power consumption of 0.4 and 3.8 watts, respectively.

TABLE 5.1: Time and energy performance characterisation obtained from DS-Explorer.

Function	Execution time (ms)	Power (W)
ReduceImage	1.299	0.4
BinariseBuffer	0.960	3.8

These performance data can then be used to provide a first realistic baseline as input to Placer. Alternatively, if an expert FPGA developer estimated that the current C code of the ReduceImage task could be improved reducing its current execution time by 10 but slightly degrading its energy performance by 2 then the expert could adjust the performance data from the table above to reflect such improvement. In our example, the performance data of the

ReduceImage task input to Placer would then be adapted from 1.299 ms to 0.1299 ms for time performance and from 0.4 W to 0.8 W for energy performance. Then the duration and energy consumption characterisation provided as input to Placer could be modified to 0.65 (duration) and 0.8 (avgpower). In the following subsection, an FPGA is added to the Aquascan example, based on time-performance characterisation provided by DS-Explorer for the longest-running DigitCorrelation function.

5.6.5   Adding an FPGA to the Hardware Platform

In a second iteration, we add an FPGA to the Placer Hardware model. Basically, a set of four identical tasks were selected, namely the DigitXCorrelation. It is a very long task. For these four tasks, a second alternate implementation is provide for executing on an FPGA (in addition to their original implementation fit for CPU execution).

DS-Explorer was used to transpose DigitXCorrelation C code to an FPGA executable kernel, then after executing it on an representative input data set, it is found that the task execution roughly divides by 5 compare to the CPU execution time.

First, we need to specify to Placer what FPGA are, what are their resources, etc. This is done below in the meta-level, and it defines a FPGA as a permanent task-processing element. It means that any number of implementations can be loaded on a given FPGA and called in parallel. The implementations loaded on a FPGA must fit within the available resources of the FPGA. The model specifies only one resource, that is the number of gates.

```
1   "permanentTasks":{ "name":"fpga",
2   "resources":["gates"],
3   "properties":[]
4   }
```

Once the notion of FPGA is added to the meta-level, we can specify implementations that target FPGAs. The Digit1Correlation can be specified to have two implementations, targeting CPU and FPGA, respectively. Placer will have to select the most appropriate one. An implementation targeting FPGA must specify the resources declared at the meta level, that is: the number of gates that it uses. The four DigitXCorreltation tasks are modified similarly. It is worth noting that the duration of the FPGA implementation is divided by 5 (from 43,205/5 = 8640) from the characterisation data obtained with DS-Explorer.

```
1  {
2  "name" : "Digit1Correlation",
3  "implementations" : [
4  {
5  "name" : "CpuDigit1Correlation",
6  "target" : "core",
7  "computationMemory" : "2",
8  "duration" : "43205"
9  }, {
10 "name" : "FpgaDigit1Correlation",
11 "target" : "fpga",
12 "resourceUsage" : [{ "name" : "gates",
13 "formula" : "300"}],
14 "computationMemory" : "2",
15 "duration" : "8640"
16 }]
17 },
```

We also add an actual FPGA to the hardware model given to Placer, with a gate resource of 1200, so it can host the four DigitXCorreltation.

```
1  {
2  "processorClass" : "fpga",
3  "name" : "Fpga1",
4  "resources" : [{ "name" : "gates",
5  "value" : 1200}],
6  "memory" : 32768
7  }
```

And the FPGA is then connected to the other cores by means of an additional (slower) bus that is slower than the on-ship bus that connects the four cores together:

```
1  "halfDuplexBus" : {
2  "name" : "slowerBus",
3  "relatedProcessors" : ["Core1", "Core2", "Core3", "Core4",
4  "Fpga1"],
5  "timeUnitPerDataUnit" : 4,
6  "latency" : 1
7  }
```

Placer delivers the following mapping that makes full usage of the FPGA to provide a global speedup.

```
 1  Mapping(makespan:47046){ Core1:1425:
 2  BarcodeResearch(dur:1425) Core2:854:
 3  Digit2PatternsIdentification(dur:854) Core3:854:
 4  Digit4PatternsIdentification(dur:854) Core4:42285: Input(dur:1),
 5  AvgFieldBrightnessCorrection(dur:934),
 6  LightingNormalisation(dur:8690), ...
 7  Output(dur:1) Fpga1:18255: Digit1Correlation(dur:8640),
 8  Digit2Correlation(dur:3205),
 9  Digit3Correlation(dur:3205),
10  Digit4Correlation(dur:3205) globalBus:8406 slowerBus:2376
11  }
```

5.7   Related Works

Many approaches have already been proposed for design-time scheduling and mapping [236]. They generally rely on ad-hoc heuristics that are hard to adapt to new dimensions. For instance, the HEFT heuristics has been proposed in [131]. Placer is about proposing an open-source tool that builds on constraint programming (CP) techniques, which have greatly improved lately. These CP techniques provide enhanced flexibility for easily and quickly adapting the mapping optimization to handle new criteria.

Some frameworks, such as PREESM, have explicitly targeted the modelling of distributed applications over embedded heterogeneous hardware architectures [308][318]. The tool allows simulating and generating code for target hardware, leveraging an S-LAM model (System Level Architecture Model). S-LAM provides a high-level architecture description through modelling of processing nodes, communications nodes, and data links. It allows easily identifying bottlenecks (latency, memory) and suboptimal resource loads. However, the tool is designed for a specific type of application that would comply with the PiSDF (Parameterized and Interfaced Synchronous Dataflow) dataflow MoC (Model of Computation). This is typically the case for dataflow-oriented signal-processing applications. Furthermore, PREESM requires the user to provide an input simulation scenario that specifies the mapping constraints, the simulation timings, and parameters, and to choose the transformations that can be applied. For code generation, the tool assembles chunks of manually-implemented code that matches the actors' functionalities as modelled in the MoC.

The company Silexica also provides the SLX tool that maps software applications onto heterogeneous MpSoC and preforms data-flow reasoning [30]. It first extracts a data-flow model from software, hence identifies opportunities for parallelization, computes a mapping of the data-flow onto hardware, and performs the mapping by generating appropriate source code. Placer, on the other hand, has a more coarse-grained approach with respect to task granularity and can select appropriate implementation and target for each task among a set of specified ones. Besides, very little information is available on the mapping algorithm used in the SLX tool.

From a FPGA perspective, Accelize company uses KPN (Kahn Process Networks) modelling to generate FPGA hardware implementations for accelerated function. Their QuickPlay solution, leverages in a seamless way, the underlying specific frameworks from FPGA chips providers (mainly Xilinx and Intel/Altera), and enables FPGA kernel generation with higher level of abstraction.

5.8   Conclusion and Future Work

Placer allows experimenting with different strategies to find a mapping and a schedule on modelled target hardware providing the availability of characterization data of application tasks for the available processing elements.

The availability of a graphical front-end is expected to significantly enhance and ease the usability of the tool specifically for visualizing the output results of the proposed mapping and scheduling. The development roadmap for Placer, until the end of the project and beyond, includes various scientific explorations towards advanced features support:

- The efficiency of the solver will be further improved. There are two opportunities for improvement, namely the CP search heuristics and the relaxation procedure used for the LNS search.

- Placer has a set of search procedures that are based on standard bricks commonly found in CP solvers. A more sophisticated one could be developed, based for instance on the HEFT heuristics or other greedy heuristics that are commonly used for scheduling [186]. It is however unclear how such heuristics could be efficiently exploited by a CP solver since they tend to make decisions based on the level of impact of the decision: decisions that are very critical are to be set first (in the good way), and other ones later; HEFT and other greedy heuristics tend to make their scheduling decisions following the flow of time.

- Several alternatives can be considered for a method that selects the assignments to relax. These should be benchmarked against a large range of instances.

- Regarding scalability, more extensive testing is needed on Placer to clearly set the size limit within which it can be used efficiently. Alternative algorithms based on local search approach could also be used, based on the OscaR.cbls framework in order to address such scalability issues [317][299].

On the other hand, as stated above, Placer needs characterization data. Such characterization data can easy be obtained for CPU/GPU targets but could be complex for FPGA. Hence feeding Placer with relevant FPGA characterization data would enable exploration of more alternatives at design time, which is beneficial for the application designer. The TANGO Design Time Characterization tool variant in year 3, DS-Explorer tool, is a recent implementation that took over the initial Poroto release. DS-Explorer aims to provide a broader FPGA design space exploration that provides higher-level abstraction, rapid evaluation, and more automated collection of characterization data. The current version of DS-Explorer provides exploration of global parameter tunings for the FPGA design. Execution time measurement is still to be automated in a generic way.

The next features to be supported by DS-Explorer are:

- Automating the deployment and execution of alternative implementation on FPGA instances at Amazon Cloud.

- Generating characterization data compliant to the Placer input format.

The tool will continue to be developed after the end of the TANGO project to provide full support for Intel/Altera targets along with current Xilinx targets.

6

Middleware, Infrastructure Management, and Self-Reconfiguration for Heterogeneous Parallel Architecture Environments

R. Kavanagh and K. Djemame

University of Leeds, Leeds, UK

CONTENTS

Hardware in various environments such as High-Performance Computing and Embedded Systems has become ever more heterogeneous in order to improve computational performance and as an aspect of managing power and energy constraints. This increase in heterogeneity requires middleware and programming model abstractions to eliminate additional complexities that it brings. This chapter explores self-adaptation including aspects such as automated configuration and deployment of applications to different heterogeneous infrastructure and for their redeployment. This therefore not only mitigates complexity but aims to take advantage of the existing hardware heterogeneity. The overall result is a self-adaptive framework that manages application Quality of Service (QoS) at runtime, which includes the automatic migration of applications between different accelerated infrastructures. Two case studies are considered: HPC and embedded systems. Discussion covers when this migration is appropriate and quantifies the likely benefits.

6.1   Introduction

In recent years considerable commercial interest in utilising heterogeneous hardware architectures (e.g., CPUs, GPUs, FPGAs) has arisen from advances in distributed computing research. The primary focus of this interest has been on improving performance and reducing overall power and energy consumption. Solving issues with energy consumption is seen as a key enabler of exascale performance [352]. Heterogeneous hardware's prevalence in High-Performance Computer (HPC) systems has led to significant complexities in scheduling work within data centres. This increased heterogeneity has given rise to the need for abstractions that simplify the use of such infrastructures, ensuring that the maximum benefit may be obtained.

The use of large-scale systems with heterogeneous resources, implies the need for multiple application implementations. This complexity gives rise for the need to maintain Quality of Service (QoS). This is a complex task, given the multiple alternative implementations of applications available and the deployment options. Each alternative mapping of application to hardware obtains different performance, power, and energy characteristics, which need to be understood within the context of a heterogeneous infrastructure, e.g., a large-scale HPC system, an embedded system, etc. of competing application deployments.

Added to the complexity derived from the heterogeneity, computer systems have faced significant power consumption challenges over the past 20 years. The dual challenge of both power and performance has in recent years shifted from the pure hardware level, to their current position as key constraints for system architects and software developers. A common theme is the need for low-power computing systems that are fully interconnected, self-aware, context-aware and self-optimising within application boundaries [221].

The need for the maintenance of Quality of Service (QoS) within complex systems gives rise for the need for self-adaptation. Self-Adaptive Systems have seen a significant level of interest in different research areas such as autonomic computing and pervasive computing [237]. They provide self-management properties and exhibit system properties such as self-awareness to achieve adaptation. They are capable of monitoring their resources, state and behaviour. Thus, QoS, power saving, performance aspects such as fast computational speed are key requirements for applications running on complex underlying hardware.

This chapter addresses these issues with a energy-aware self-adaptive framework for heterogeneous parallel architectures. The main contributions of this chapter are:

- a framework for managing heterogeneous hardware and optimising deployments through self-adaptation

- an event-driven self-adaptation manager for runtime-based adaptations

- recommendations and analysis on comparisons of deployment solutions on heterogeneous devices considering two case studies: HPC and embedded systems.

The remainder of the chapter is organised as follows. In Section 6.2 we present the overall architecture that supports energy awareness and self-adaptation. Section 6.3 gives a detailed discussion of adaptation within the proposed framework. In Section 6.4 the experimental setup is discussed, followed by experimentation and evaluation in Section 6.5. The related work is then presented in Section 6.6, and Section 6.7 summarises the research and provides plans for future work.

6.2   Architecture

Given the importance of heterogeneous hardware environments and the need to maintain both QoS and lower power consumption, the architecture shown in Figure 6.1 is proposed. It includes the high-level interactions of all components, separated into three distinct layers and follows the standard application deployment model. Its aim is to control and abstract underlying heterogeneous hardware architectures, configurations and software systems, while providing tools to optimize various dimensions of software design and operations (energy efficiency, performance, data movement and location, cost, time-criticality, security, dependability on target architectures). Next, details on the interactions of the architectural components are discussed.

6.2.1   Layer 1—Integrated Development Environment (IDE)

The first block, the IDE layer facilitates the modelling, design and construction of applications, in order to aid the evaluation of power consumption. A number of IDE plug-ins are provided as a means for developers to interact with components within this layer. Lastly, this layer enables architecture agnostic deployment of the constructed applications, while also maintaining low power consumption awareness. The components in this block are:

Requirements and Design Tooling: guides the development and configuration of applications to determine what can be targeted in terms of both Quality of Service (QoS) and power consumption when exploiting the potential of heterogeneous hardware devices;

Programming model (PM): supports developers when coding their applications sequentially by letting them annotate their programs in such a way that the PM runtime can then execute them in parallel on heterogeneous parallel architectures being aware of the power consumption. The PM is implemented as an hierarchical combination of the COMPSs [54] and OmpSs [137] task-based programming models, where COMPSs manages the application in the distributed platform and OmpSs inside each compute node.

Code Profiler: This achieves power reduction through the the software development process by providing developers the ability to directly understand the energy footprint of the code they write.

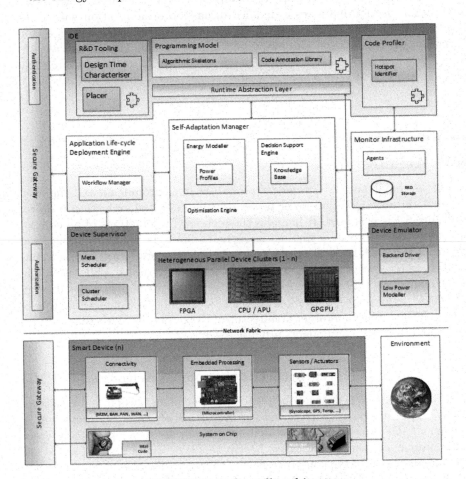

FIGURE 6.1: Overall architecture.

6.2.2   Layer 2—Middleware

The second block handles the placement of an application considering energy models on target heterogeneous parallel architectures. Its tools are able to assess and predict performance and energy consumption of an application. Application-level monitoring is also accommodated, in addition to support of self-adaptation for the purpose of making decisions using application level objectives. The components in this block are:

Application Life-cycle Deployment Engine (ALDE): this component manages the life cycle of an application deployed by the IDE. The ALDE introduces four entities for each application: Application, Executable, Deployment, and Execution Configuration. More details can be found in Section 6.3.1.

Monitor Infrastructure (MI): provides monitoring of the heterogeneous parallel devices (CPU, memory, network ...) along with historical statistics for device metrics. The monitoring of an application includes power, energy consumed, and performance.

Self-Adaptation Manager (SAM): This component provides key functionality to manage the runtime-based adaptation strategy applied to applications and Heterogeneous Parallel Devices (HPDs). This includes aspects such as initiating redeployment to another HPD, restructuring a workflow task graph or dynamic recompilation. Furthermore, the component provides functionality to guide the deployment of an application to a specific HPD through predictive energy modelling capabilities and polices. Further details can be seen in Section 6.3.2.

6.2.3   Layer 3—Heterogeneous Devices

The last block, addresses the heterogeneous parallel devices and their management. The application admission, allocation and management of HPDs are performed through the orchestration of a number of components. Power consumption is monitored, estimated and optimized using translated application-level metrics. These metrics are gathered via a monitoring infrastructure and a number of software probes. At runtime HPDs will be continually monitored to give continuous feedback to the Self-Adaptation Manager. This ensures the architecture adapts to changes in the current environment including energy demand. The components in this block are:

Device Supervisor (DS): provides scheduling capabilities across devices during application deployment and operation. The component essentially realises abstract workload graphs, provided to it by the Application Life-cycle Deployment Engine, by mapping tasks to appropriate HPDs.

Device Emulator: provides the initial mapping of the application tasks onto the nodes/cores (at compile time). The mapping procedure is static and thus does not take into account any runtime constraints or runtime task mapping decisions.

Secure Gateway: supports pervasive authentication and authorization, which is at the core of the proposed architecture enables both mobility and dynamic security.

6.3   Adaptation Framework

This section is comprised of two main subsections; the first subsection 6.3.1 discusses enabling technologies for self-adaptation with heterogeneous architectures. This focuses in particular on the application life-cycle management and how this enables multiple deployment configurations can be constructed. After this runtime considerations are considered with the Self-Adaptation manager (Section 6.3.2) and how the adaptation can be guided by an energy model.

6.3.1   Adaptation Enablement

In order to utilise heterogeneity it is generally required to have an application written to target the specific infrastructure though code specifically compiled to target the device. This gives rise to either modules of an application or a whole application having to target a specific infrastructure or target instruction set. There is the potential to use multiple implementations or configurations of an application therefore to allow for heterogeneous hardware to be fully utilised. Once the middleware can be made aware of these implementations, adaptation considering heterogeneity can be realised.

One such way of realising these implementation differences is through the ALDE. An *Application* can be built by the ALDE and optimized for different heterogeneous architectures. Each application can have a series of *Executables* which are optimized for different scenarios, such as CPU computation only or CPU+GPU, etc. The ALDE for a given system can upload all the necessary executable files to that platform creating a *Deployment*. Each deployment in turn can have several associated *Execution Configurations*, which indicate which heterogeneous resources the Deployment needs. In some environments the ALDE may be unavailable, in this case it is possible through the programming model runtime to make the different deployment options available.

These configuration options can then be ranked by the IDE tooling, or through benchmarking and through examining the environment the best available deployment can be utilised. This ranking is used in the case of the ALDE to make the best deployment decision and by the Self-Adaptation Manager in cases of reconfiguration.

6.3.2 Runtime Self-Adaptation Management

The Self-Adaptation manager's (SAM) principle role is to manage application-level adaptation at runtime, managing trade-offs between energy, power and performance within the framework. It is event-driven, deciding for each event what adaptation to take and where it should be applied.

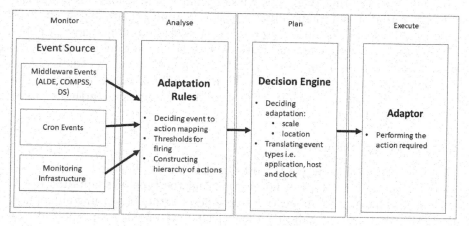

FIGURE 6.2: The flow for self-adaptation.

It works through a series of listeners that monitor the physical infrastructure, the middleware and the system clock for cron-based events (see: Figure 6.2). The listeners act as triggers generating events which are then mapped to actuators that perform the required adaptation. Even with a mapping between the event and actuator a decision needs to be made in regards to the scale of adaptation. Events can be caused by applications, physical hosts or cron events; these different event types have different information associated with them, which needs translating into the appropriate action.

6.3.2.1 Event Generation

The first step in adaptation is a notification event which derives from the listeners. These events principally contain the following information:

Time: the timestamp of the event.

Value: a raw measured value representing the scale of the QoS breach.

Severity Type: i.e., "violation", "warning", or "informative".

Agreement Term: the metric to be monitored.

Guarantee Id: an identifier for each QoS constraint.

Operator: such as greater than, less than, equal.

Guaranteed Value: the value of the threshold.

Events (Host, Application and Clock) dependant upon their source must contain additional information. *Host* events additionally must contain the hostname, thus indicating the events' origin. *Application*-based events must additionally record the application's name, a reference to the application instance and a reference to any application configuration information and application specific firing rules. *Clock* events, must hold a map for additional settings so that it can mimic either a host or a application-based event. This allows clock events generated by previous rules firing to mimic host or application based events, facilitating features such as un-pausing an application, or rebooting a host. Examples of events might be: boundary conditions on measurements, host state changes (idle hosts, failing, failed, or draining of work), applications approaching their deadline, application starting/completion, cron-based events.

6.3.2.2   Adaptation Rules

On event notification the SAM works in *two phases*. The first considers the mapping between the type of notification and the type of adaptation to use such as: redeploying an application to use accelerators or pausing an application. The second phase indicates the exact nature of this adaptation to take such as which application should be adapted and by how much. The first phases mapping utilises rules that can be specified as a tuple of:
⟨Agreement Term, Comparator, Response Type, {Event Type}, {Lower Bound}, {Upper Bound}, {Parameters} ⟩   which is utilised to determine the form of adaptation to take. Two examples of this are:

⟨IDLE_HOST+ACCELERATED,EQ, RESELECT_ACCELERATORS⟩ and
⟨IDLE_HOST+ACCELERATED, EQ, RESELECT_ACCELERATORS,
WARNING, 0, 0, KILL_PREVIOUS=TRUE; application=gromacs⟩.

The latter optional values allow for stronger granularity ensuring the adaptation behaviour considers the scale of the notification event. This provides the flexibility to do things such as: i) Responding to warnings, in a different fashion to breaches or informative notifications. ii) Observing the difference between the guaranteed value and the measured value and providing a stronger response if the deviation is further away (i.e. the lower bound and upper bound values). iii) Parametrising the rules, so applications can further indicate how adaptation should occur, e.g., clock-based events such as "it is out of working hours" can specify through parameters application information, thus allowing lower priority jobs to run.

Events once triggered need not always cause adaptation to occur. Application-and resource-based events, derived from measurements utilise a event count threshold value, which ensures that the temporary reporting of minor breaches can be ignored (e.g., if power consumption goes too high due to a short burst of CPU utilisation). Additionally once a rule has fired, a recent history log prevents the same rule firing in rapid succession, thus avoiding over-adaptation. After a short time, the rule can then be re-fired. Rules can optionally be set into a hierarchy so that if one cannot be applied additional rules that match the criteria may be used instead. This generates the prospect of fall-back options or an intensification of the adaptation response.

6.3.2.3   Scale of Adaptation

The second phase decides upon the location and scale of adaptation. This involves the usage of a decision engine that considers various parameters, such as the application configuration, QoS goals (e.g., save energy, cap power, reduce completion time) and the current environment to decide where the adaptation should take place. An example of this might be while scaling up resources, ensuring no breach of other criteria such as a power cap occurs.

The decision engines handle cases where information is lacking on what to adapt. This includes cases such as host-based events and their transformation into actions applied to applications. The transformation process for host-based events can be achieved for example: *randomly*, based upon the applications' power consumption, or based upon the *last application instantiated* on the originating host.

Clock events can be transformed into either host or application based events dependant upon the additional parameters attached to the event. They are transformed in order of precedence by: Event data has *application* details attached, which might, for example, occur when a pause action has specified when to resume. Event data has *host* details attached (such as when a shutdown event has a reboot option) and finally when the decision rule contains the *host* or *application* data.

6.3.2.4   Actuation

The actuation can cover many different forms of change. This can be at the level of applications/workload, for example: increasing/reducing an application's wall time, adjusting the wall time so it's closer to the runtime, aiding scheduling and in particular backfilling, pausing and restarting applications or sets of applications, oversubscribing applications to resources, reselecting the accelerators an application uses, killing off an application, or it can be at host level such as starting and shutting down hosts and adjusting the cluster's power cap.

One actuator stands out as being more complex than the rest, "Reselect Accelerators" (see Algorithm 3). Its primary aim is to choose an application configuration that is better than the existing configuration. This may, for example, be switching from a single threaded CPU-bound executable to a GPU-accelerated version of the same application. This could be done to improve the accelerator utilisation. The algorithm first filters out configurations that are already running and thus have a head start upon any new instance starting. The second phase in the algorithm selects the new instance to launch. This works by ranking each configuration by either, power, energy or completion time. The best configuration is then selected so long as its power/energy or completion time is better than the existing running configuration. This ranking is performed based upon pilot jobs that are executed beforehand. The process generating the ranking is as follows:

1. launch fixed-workload pilot jobs: (either a single representative value, or uniform range, applied to each alternative configuration).
2. record the following information (application_id, job_number, completion time, energy consumption, configuration_id).
3. for a given job filter out configurations that cannot run and calculate the average power, energy consumption and completion time for a given application_id and configuration_id.
4. order by chosen ranking (completion time, power or energy).

The pilot jobs have a fixed workload (or a sequence of workloads that is repeated uniformly against each configuration), ensuring that each application configuration is compared fairly. This comparison gives a relative ranking between the configurations based upon the current hardware setup. It is considered that each application configuration has a relative affinity to each of the available resources on the testbed, therefore if the pilot jobs are repeated several times the likely improvement between configurations is going to be realised. An example of this affinity is where CUDA-compiled applications must be launched upon a GPU-enabled node, thus constraining a configuration to launch on a subset of the total amount of nodes available, which narrows the likely range of possible power, energy and completion time values for the fixed workload on the testbed. This process enables the ratio of improvement between the configurations to be determined. This includes aspects such as the likely energy consumption and average power consumption for running a pilot job or job (by relative ratio between configurations), which reflects complex aspects such as which resources a particular job was submitted to.

6.4 Experimental Design

To evaluate the feasibility of the adaptation features as outlined in Section 6.3, two sample use cases are utilised to show self-adaptation within different contexts. Experiments are designed in the context of the energy-efficient HPC environment presented in Section 6.2 as implemented by the TANGO project [221].

The objective of the experimentation is to ascertain if the self-adaptation when monitoring applications in operation achieves dynamic energy management from the middleware of a HPC software stack. First an outline of the experiments is given followed by the testbed and application setup.

Algorithm 3 Reselection of Accelerators

---

procedure RESELECTACCELERATORS(String appName, String deploymentId, boolean killPreviousApp, RankCriteria rankBy)

    AppConfig currentConfig ← getCurrentConfigurationInUse(appName, deploymentId)

    AppDefintion appDef ← getApplicationDefintion(currentConfig)

    AppConfig validConfigs[] ← getValidConfigurations(appDef)   ▷ check resources availability and executables are compiled

    validConfigs[] ← removeAlreadyRunningConfigurations(validConfigs[])

    AppConfig selectedConfig ← selectConfig(validConfigs[], appDef, currentConfig, rankBy)

    startAndStopNewAndOldJobs(selectedConfig, appDef, currentConfiguration)

end procedure

procedure SELECTCONFIG(AppConfig validConfigs[], AppDefintion appDef, AppConfig currentConfig, RankCriteria rankBy) sort(validConfigs[],rankBy)

    if first(validConfigs[]).isRunning() **then**

        return null

    end if

    if isRankBetter(first(validConfigs[]),currentConfig) **then**

        return first(validConfigs[])

    end if

    return null   ▷ no better solution so return

end procedure

The experiments center around two sample use cases. In the first case we see the prospect of hardware becoming available; this might be the completion of another job for example. This presents the opportunity to trigger adaptation and redeploy an application in order to obtain an improvement and may include changing accelerators available. A similar scenario is the loss of resources, such as a node failure, whereby the reselection process for jobs may be required.

In the second scenario a power-smoothing mechanism for the data center is implemented, at application level. Time-based rules for each application are used to ensure a low-priority application is paused during a "busy time of the day" and then resumed later. Thus smoothing power consumption.

The experiments were performed on a cluster, using a subset of a bullx blade system. The testbed was composed of the following heterogeneous hardware resources.

4 bullx 515 nodes equipped with: 2 Intel Xeon E5-2470 (Sandy Bridge) at 2.3GHz, 12 × 16GB DDR3-1600 ECC SDRAM and 2 × 256GB 2.5″ SATA3 flash SSDs. Additionally two of these bullx 515 nodes (ns50-51) have 2 Nvidia Kepler K20X GPUs each and the other two bullx515 nodes (ns52-53) have 2 Intel Xeon Phi 5100 2 Intel Xeon Phi 5100 series (rev 11) KNC each.

In addition to these nodes there are: 3 bullx B520 double compute blades (ns55-57), each equipped with: 2 Intel Xeon E5-2690 v3 (Haswell) at 2.6GHz with 16 × 16GB DDR4-RDIMM 2133DDR and 2 × 256GB 2.5″ SATA3 flash SSDs.

The experiments utilise the GROMACS (http://www.gromacs.org/) application, an open source and widely-utilised molecular dynamics simulation package. It is used to generate load within the testbed and provides a realistic application that can be compiled into various alternative implementations such as Message Passing Interface (MPI) and CUDA.

6.5   Evaluation

6.5.1   HPC Environment

The following section discusses the performance of the self-adaptation presenting an analysis of the experimental results. The first experiment presented illustrates the use of multiple application implementations of HPC applications in order to deploy, monitor and adapt an application so that the most efficient implementation is executed, given the resources that are available at the time of execution. The workflow is as follows:

1. The Gromacs application is defined in the ALDE and the configurations available
2. Job Deployment, a CUDA instance is started

3. Launch Job, the device supervisor launches the job onto the infrastructure

4. The SAM receives an event indicating a host has become free with an accelerator

5. The SAM compares Gromacs implementations and re-launches the most efficient version of the application

6. The Gromacs application completes

In the following experiment two different versions of the Gromacs mini-app are prepared one using MPI (using 16 threads) and the other CUDA. For each configuration, previous pilot runs are performed where execution time and energy consumption are measured for a given fixed workload. This presents a relative ranking of each configuration of the application upon the available hardware. This ranking of each application configuration will have an affinity towards a particular set of resources. A CUDA job, for example, must be launched on a subset of resources that have GPUs. Table 6.1 shows these initial measurements, where we can see that depending on the configuration, the Gromacs simulations can achieve different performance and energy consumption. The most efficient configuration in terms of time and energy for this execution is using the MPI implementation of Gromacs. The variance in execution time and energy consumption is low, in this case, making it particularly suitable for determining speed-up between configurations.

TABLE 6.1: Run of pilot jobs to determine power, energy, and completion time rankings.

Name	Run Count	Total Energy (all runs)(J)	Average Energy (J)	Total Time	Average Time Per Run (s)	Average power (W)
cuda	3	22,835	7,611.67	87	29	**262.47**
mpi 16	3	**19,217**	**6,405.67**	**67**	**22.33**	286.82

What can be seen is that the MPI application has the lowest energy consumption overall at 6,405.67J per run. This is principally due to the lower runtime of the MPI application, 67s as compared to 87s. The average power consumption of the CUDA application is, however, less. This set of configurations therefore offers either:

- A lower power consumption that runs for longer and consumes more energy.
- A higher power consumption that runs for a shorter period of time and therefore uses less energy.

These results give an indication of how quickly a replacement replica should start in order to consume less energy overall, assuming the overall application workload is similar to that of the pilot jobs. To find the relative rank between configuration options requires each application configuration to remain proportional in its energy usage to the other potential configurations.

In the case of the pilot job executions shown above, an MPI implementation could be started in the first 4s of the CUDA instance's execution and still use less energy. This is derived from:

avg(cuda_run_energy) - avg(mpi_run_energy) = Δenergy_between_run_types
7611.67 - 6405.67 = 1206J
Δenergy_between_run_types / avg(mpi_run_power) = migration exploitation window size
1206 / 286.82 = 4.02s

Given the executions above are short-lived the benefits of restarting jobs with different accelerators is limited; however, if the workload is increased then the migration exploitation window size will also increase. Practically as this mechanism only provides the relative ranking between application configurations, without a priori knowledge of the runtime (to work out scale differences from the pilot jobs), then it is possible to use recent job submissions as guidance on current expected execution durations. The MPI and CUDA jobs both linearly scale (so far as tested), leaving the ratios between their durations and energy consumption comparable. However, due to gradient differences causing divergence (Figure 6.3), CUDA increasingly uses more energy in comparison to the MPI implementation; therefore to save both energy and time, larger jobs should favour the MPI implementation. The migration exploitation window size scales linearly with the job size.

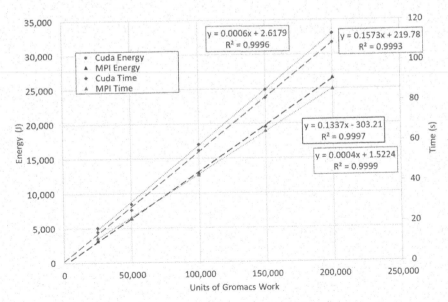

FIGURE 6.3: CUDA vs MPI job workload scaling.

In addition, applications using this calculation may be ranked to consider which ones have the most difference between the available configurations options. This therefore finds the application which is most likely to benefit from adaptation.

To illustrate this a rule is added that causes the detection of idle resources with accelerators. The SAM therefore considers if any Gromacs instance may be accelerated. The rule used to generate the migration of the application is shown in Table 6.2.

TABLE 6.2: Migration rule.

Agreement Term	Comparator	Response Type	Event Type	Parameters
IDLE HOST + ACCELERATED	EQ	RESELECT ACCELERATORS	WARNING	KILL PREVIOUS=TRUE; application=gromacs

The reselect accelerators actuator compared the deployment options available and found it was possible to execute an instance that consumes a lower amount of energy. The sequence of events generated is therefore that an IDLE HOST ACCELERATED event triggers the adaptation, which terminates the CUDA instance and launches a new MPI instance. The result of this is shown in Figure 6.4. The CUDA job runs on node ns51 and is then replaced by an MPI job on node ns55. Followed by the MPI job then completing at 27s, 2 seconds earlier than the CUDA job could have been expected to complete. Although this is a small benefit of 6%, with longer-running jobs the benefit is likely to be more substantial.

This job redeployment mechanism allows for multiple possibilities, when considering events that the SAM will act upon. For example, job or node failure could cause the job redeployment to the most appropriate available resources, thus acting as a recovery mechanism. If ranking by power consumption it would aid power-capping mechanisms by ensuring the mix of jobs running is most likely to have a sufficiently low power consumption. This mechanism also ensures accelerators where appropriate are more likely to be utilized.

In Figure 6.3 it seems counter-intuitive that the CUDA implementation, given the vast parallelism of the GPU, does not perform better than the MPI implementation. This results in the transfer of the job from ns51 to ns55 in Figure 6.4. The difference is caused by the overall efficiency of the host. The explanation can be seen in Figure 6.5, which shows the energy consumption of a series of Gromacs jobs, both on ns51 (default choice for GPUs) and ns55 (default choice for MPI jobs), along with MPI implementation on ns51. We omit here the graph for job completion time for purposes of clarity though it is similar to Figure 6.3 and Figure 6.5.

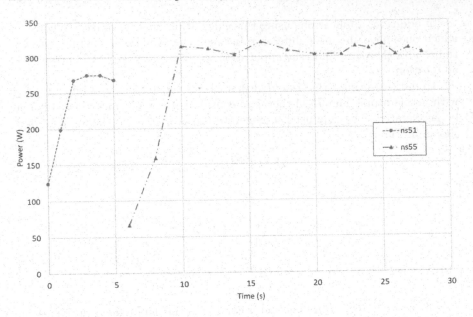

FIGURE 6.4: MPI job launching and cancelling CUDA job.

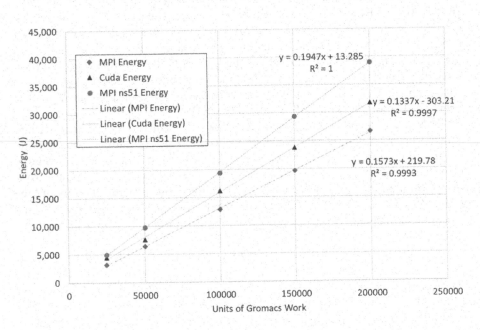

FIGURE 6.5: MPI vs. CUDA on multiple host types.

The CUDA implementation is more efficient than the MPI implementation of ns51, however the MPI implementation may also run on ns55, which is generally more efficient than ns51 due to CPU generational differences, such that the MPI implementation is the best choice should the more efficient nodes be available.

The strategy employed here to determine the relative ranking of jobs has many considerations that must be made, which are going to be split into categories and discussed in turn. In terms of comparing various different configurations of an application then there are two key aspects: temporal and power.

Job duration is important, simply the longer it runs the greater the power consumption. However ranking of jobs may be subject to changes in throughput over time for a given configuration, dependent upon the size of the workload (Figure 6.3 and Figure 6.5). This can be checked with a series of jobs of different sizes and examining if the application scales under workload changes. If an application scales in a predictable fashion, recently completed jobs can be used as a guide to estimate both completion time and the migration exploitation window size. Thus offering an estimate of the likely speed-up/energy-saving between configurations. One strategy of tackling throughput changes would be to make pilot jobs of a similar size to "typical" jobs thus mitigating scaling differences.

The second aspect is the power consumption of a job and if its power consumption is consistent (with obvious exception of the start and end of jobs). In the case of Gromacs (MPI) on ns55 the power consumption is consistent in the range of 303-320W with an average of 310W excluding earlier lower-power consumption at the start of the applications' execution. If it varies through time, the following questions can be considered. Does it act in phases with any particular recognisable fashion? And more importantly does the average power consumption stay the same with job length?

Applications may be made up of several stages. These from the perspective of ranking configurations only need be considered in cases where the average power consumption varies or the duration varies because of some underlying change in actions. Examples of this might include transferring data vs compute work, or different types of compute work on different accelerators/instruction sets. The steps may change size, altering average power consumption (e.g., in cases where a larger input file is needed). Data transfers may alternatively act as bounding behaviour on the power consumption, for example streaming where only a certain amount of power consumption can be achieved during transfer.

An additional aspect is the heterogeneity of the hardware. An applications configuration is unlikely to specify specific hardware, but it may provide an affinity to a given subset of hardware or explicitly exclude some resources. An example of this is a CUDA job that needs a GPU. It may be for a given job configuration that the resources available to use are not very heterogeneous in terms of power and compute capacity (i.e., temporal/duration of job given the same workload). It is likely that this affinity of a given configuration towards different subsets of the infrastructure can be discovered by several runs of a given pilot job. Thus giving an average case for the likely speed-up between configurations can be discovered.

This strategy has its limitations and is dependent upon other workloads running at the same time, as well as the scheduler in use and size of the job to be rescheduled. The variance in time and energy consumption can be considered a measure of the reliability of the reselection method. The variance would be reduced if the pilot run dataset could be filtered to records only considering the types of resources available.

6.5.2   Remote Use Case Environment

The second experiment examines the use of the SAM in an environment that is suitable for the remote use-case environment, where it directly integrates into the programming model runtime. This importantly does not make use of either the ALDE or the infrastructure manager, which are expected to be used in HPC environments. In this case the workflow is as follows:

1. The emulated remote use-case application is launched.
2. The SAM receives an event indicating scaling should occur.
3. The SAM adds to the programming model runtime the additional resources.
4. The emulated remote use-case application completes.

The rule that generates the add-task command is called increased_workload, which is then triggered by the SAM's rest interface using curl as part of the experimentation. A scaling event that adds more compss tasks to execute the work might for example be triggered by: low frame rates, other levels of poor performance, increases in the workload such as from new users, or indications that the executable running is nearing a deadline (see also Table 6.3).

TABLE 6.3: Remote use-case scaling.

Agreement Term	Comparator	Response Type	Event Type
Increased_workload	EQ	ADD_TASK	SLA_BREACH

The result of this workflow is shown in Figure 6.6, whereby the rate of jobs being completed initially is at a given rate, followed by a scaling event which causes the rate to increase until no further work is available to be computed. In the experiment presented we focus upon a worker that uses the CPU although workers for both CPUs and GPUs exist.

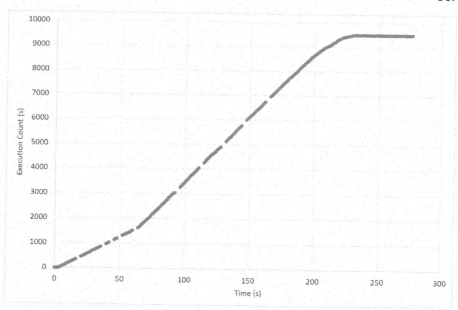

FIGURE 6.6: Emulated remote use-case scaling.

The framerate is further illustrated in Figure 6.7, where we see the framerate increase at around 60 seconds, into the trace.

TABLE 6.4: Scaling in remote usecase.

	Average Framerate (f/s)	Std Deviation	Min	Max	Range
Pre-scaling	26.62	4.26	14	32	18
Post-scaling	51.34	10.89	24	65	41

From Figure 6.6 and Figure 6.7 we can derive Table 6.4, where we can see the application has responded positively to the adaptation event, with almost linear scaling of the chosen application. Table 6.4 also illustrates that the emulation of the remote use-case application can suffer from some variance in the framerate, which is not expected to be seen outside of the emulation of the use case.

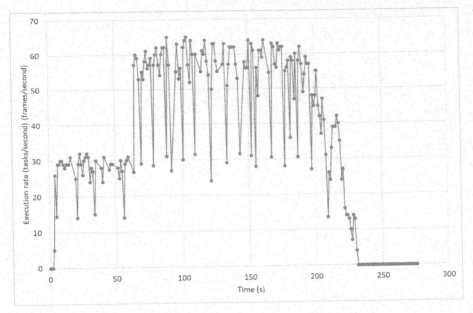

FIGURE 6.7: Emulated remote use-case frame rate.

6.6 Related Work

Research effort has targeted energy-efficiency support at various stages of the application service life cycle (construction, deployment, operation). Additionally heterogeneity has been recognized as a viable solution to improve performance and reduce power consumption at the same time [117]. This section reviews existing work and categorises it into self-adaptation, energy efficiency and heterogeneous computing environments.

Self-adaptive systems (SASs) offer solutions to address issues such as maintenance, configuration and Quality of Service (QoS) control in complicated environments. The use of self-adaptation however requires the answering of fundamental questions of "When to adapt?", "Why do we have to adapt?", "Where do we have to implement change?", "What kind of change is needed?", "Who has to perform the adaptation?" and "How is the adaptation performed?" [331].

Krupitzer et al. [237] presents a taxonomy of self-adaptive systems and their inspiration, while R. deLemos et al. [126] identifies research challenges when developing, deploying and managing self-adaptive software systems. These challenges result from the dynamic nature of self-adaptation, which brings uncertainty. Adaptation commonly follows a Monitor Analyses Plan and Execute (MAPE) approach. An extended architecture of the MAPE-K loop as a reference model for the design of self-adaptive systems is found in

[234], assuming that the system has a central controller with a central MAPE-K loop. The proposal consists in continuously evaluating adaptation steps concerning their actual effect and adaptation mechanisms concerning their applicability and efficiency in the case of topology changes. An agent-based modelling approach for adaptation is presented in [218]. An Agent Verification Engine (AVE), which constructs agents to perceive, react and adapt to runtime changes of a component-based system, is proposed. These agents are based on the Belief-Desire-Intention (BDI) architecture, in which agents operate in terms of motivation and beliefs. The work in [204] consolidates design knowledge of self-adaptive systems. To support software designers, the paper contributes with a set of formally specified MAPE-K templates that encode design expertise for a family of self-adaptive systems. The templates comprise: (1) behaviour-specification templates for modelling the different components of a MAPE-K feedback loop (based on networks of timed automata), and (2) property-specification templates that support verification of the correctness of the adaptation behaviours (based on timed computation tree logic).

One possible goal of self-adaptation is the maintenance of energy efficiency and the management of low-power environments. Kong and Liu [235] group green-energy-aware into four distinct categories, namely: (1) Green-energy-aware workload scheduling. (2) Green-energy-aware Virtual Machine (VM) management. (3) Green-energy-aware energy capacity planning. (4) Interdisciplinary. The first three areas are focusing upon workloads' temporal and spatial flexibility, VM scheduling for a similar effect and crafting workload profiles to desirable power consumption profiles. This chapter considers not only workload placement but also the exact configuration of the application, which has multiple configurations available and various possibilities to use different hardware accelerators. The EcoScale project [201, 193] uses hardware-performance monitors and models to project runtime and power consumption in heterogeneous environment using accelerators. The aim is to enable a runtime scheduler called the execution engine, to dynamically select and distribute software functions to either hardware acceleration or to software-based execution. This selection is based upon workers in the vicinity and implementing worker queues [201].

Research effort has focused on the exploitation of hardware accelerators in Cloud-computing environments by addressing the challenge of programming such systems and making them easily accessible in a virtualized environment. One common approach is to propose methods to offload computations on heterogeneous hardware components. A solution for the efficient exploitation of specialised computing resources of a heterogeneous system is found in [139]. Other works have proposed heterogeneous architectures that combine high-performance and low-power servers in order to achieve better overall energy proportionality and energy efficiency [310]. The mapping problem between compute resources and application configurations is explored in [274], considering throughput in the context of Cloud. This is similar in idea to the work presented here, though the exact context and approach differs regarding the heterogeneity that is being utilised. The ASCETiC project [130] holds similarities in that it worked upon energy efficiency and software adaptation but in the context of Clouds. A highly scalable model for developing applications, exploiting hardware heterogeneity in Cloud data centres while at the same time considering the aspect of energy efficiency is presented in [329]. In this model applications are expressed as interconnected microservices which are automatically scheduled for execution on the most suitable heterogeneous computing elements. This leaves the open problem of handling of applications that do not follow a microservices pattern.

The Legato project [246] identifies power as a key concern and while software-stack support for heterogeneity is relatively well developed for performance, is seen to remain an open question for power and energy-efficiency, which is an aspect that this chapter contributes towards. StarPU [50] is one such early work that uses abstractions to allow workloads to be placed upon various different accelerator based platforms, selecting between various different accelerators, to improve computational performance and not energy. The Antarex project [346, 345] focuses on energy-efficient systems and in particular on producing a tool chain that tunes code to run efficiently on heterogeneous infrastructures. The Manago project [155] is equally similar to work presented in this chapter but with a particular focus on time-predictability with trade-offs with power and energy efficiency. Dutot et al. [140] advance upon power-capping and consider energy budgets with a principle focus on scheduling.

From a technology viewpoint, hardware accelerators, such as GPGPUs and FPGAs still need the use of power-reduction techniques such as Dynamic Voltage and Frequency Scaling (DVFS) and partial reconfiguration for FPGAs to keep power consumption under control [283]. Many approaches to adaptation and energy/power optimisation concentrate at the hardware level, such as utilising task scheduling coupled with GPU-specific DVFS and dynamic resource sleep (DRS) mechanisms, as a means to minimise the total energy consumption [272]. The work presented in this chapter complements such hardware-based strategies given the similarity of goals, yet utilises software-based approaches to minimise power and conserve energy.

6.7   Conclusion

Future middleware systems will increasingly need to abstract away the complexities of heterogeneous hardware whilst also supporting power and energy awareness through automatic reconfiguration at both hardware and application/workload levels. This chapter has presented an overview of an energy-aware self-adaptive framework for heterogeneous parallel architectures, that aims to achieve such a goal.

A key enabler for addressing heterogeneity is the ability to reason with multiple different implementations of the same application that are prepared for the different deployment alternatives. This works across multiple environments such as HPC or remote processing use cases, each presenting their own constraints while keeping to the same generic requirements for an adaptive middleware.

This adaptive middleware has been shown a self-adaptive approach for heterogeneous hardware, including more specific aspects such automatic redeployment of applications auto-scaling and power capping. This chapter has discussed the criteria that would need to be considered as an element of the migration process. In a wider context of self-adaptation, mechanisms have been demonstrated that can aid power-capping behaviour and raise the intelligence of such an approach from the hardware to application level. Such adaptive systems are likely to become more prevalent extending into other domains such as embedded systems, as well as becoming increasingly complex with aspects such as check-pointing-based schemes as part of migration and utilising application models for ranking application deployment alternatives during migration.

7

A Novel Framework for Utilising Multi-FPGAs in HPC Systems

K. Georgopoulos, K. Bakanov, I. Mavroidis, and I. Papaefstathiou
Telecommunication Systems Institute (TSI) – Technical University of Crete, Crete, Greece

A. Ioannou and P. Malakonakis
Technical University of Crete, Crete, Greece

K. Pham and D. Koch
The University of Manchester, Manchester, UK

L. Lavagno
Politecnico di Torino, Torino, Italy

CONTENTS

This chapter presents the technology developed in the context of the European project ECOSCALE. This project aims at contributing to the effort for a state-of-the-art platform that can serve applications which can only be satisfactorily addressed in the context of High-Performance Computing (HPC). ECOSCALE is a platform that comprises of both a novel software framework that allows users to introduce their applications described in OpenCL (or any other high-level equivalent) as well as an architecture that utilises reconfigurable hardware in order to execute at high speeds computationally-intensive tasks. Therefore, the ECOSCALE platform comprises of a framework including a number of functional and technological layers, such as the hardware architecture and the middleware. This chapter will describe all key ingredients that make up this platform and in the process explain how any prospect application could be applied to it.

At a physical level, ECOSCALE introduces the Worker that is the processing unit of its technology. The Worker is effectively a single ECOSCALE FPGA featuring all elements required for a single computational node setup, i.e., a processing system, memory and programmable logic otherwise known as reconfigurable logic. In the context of ECOSCALE, the computationally-intensive tasks are meant to be implemented within the reconfigurable section of the FPGA in what is known in ECOSCALE nomenclature as *accelerator* modules. Furthermore, the complete ECOSCALE system is going to be comprised of a number of Workers and the project prototype is going to be comprised of 64 FPGAs, which thanks to ECOSCALE's UNILOGIC technology, will make all reconfigurable hardware available for exploitation by any given task or application in a seamless and transparent manner. Note that an added benefit from using FPGA technology is the potential for a low-power HPC system.

A typical ECOSCALE flow will commence with a hypervisor that administers all incoming task requests for execution on the ECOSCALE platform. Based on resource utilisation information, it will then allocate these tasks onto the reconfigurable resources for execution. Accelerator modules will then be either directly fetched from the ECOSCALE accelerator library and deployed onto the FPGA reconfigurable fabric or automatically generated by the ECOSCALE accelerator synthesis tool and then added to the library for execution on the reconfigurable logic. Consequently, accelerator modules are downloaded onto the reconfigurable hardware resources and process data stored on a shared global address space. This allows for massive parallelisation and consequently significantly fast execution times that set the foundation in the effort for exascale processing in ECOSCALE.

7.1 Introduction

High-Performance Computing (HPC) lays at the heart of many modern technological challenges and milestones. It has been the unavoidable outcome of the evolution towards contemporary scientific analysis and study, and it relates to a number of crucial fields, e.g., aerodynamics, quantum mechanics, oil and gas exploration and many other, that have a significant socio-economic impact. HPC is primarily associated with highly computationally-intensive applications and the metric commonly used for quantifying HPC platforms is the number of floating-point operations that a system can achieve over a finite amount of time, usually a second.

The range of technological approaches that can serve computationally-intensive applications in the context of HPC systems has been wide and modern HPC computing systems have managed to achieve a performance in the order of petaFLOPS, that is 10^{15} floating-point operations per second. Example top systems related to this performance metric are the *Summit* [23] in the United States with a score of 122.3 petaFLOPS and the *Sunway TaihuLight* [135] from China that achieves 93 petaFLOPS.

Initial supercomputing efforts focused on stronger processing capabilities, however, in the 1970s parallelism came into the picture and significantly boosted the performance capabilities for such systems. HPC systems rely on highly effective architectures combined with powerful processing elements. These processing elements have also varied and the main representatives are the CPU, General-Purpose GPU (GPGPU), Field-Programmable Gate Array (FPGA) and dedicated hardware, i.e., ASICs. In fact, the use of reconfigurable hardware in the form of FPGAs for HPC systems is a fairly recent innovation, and this technology constitutes the main processing element of the system and subsequent framework proposed within the domain of EU-funded project ECOSCALE [5]. As a solidifier to the validity of such endeavours comes Microsoft's Catapult project [18] that uses reconfiguration to further advance Cloud computing, which demonstrates the interest of major corporations for the potential that FPGA technology holds.

ECOSCALE's role in the context of high-performance computing is important primarily due to the fact that the target of its scientific advancement is the production of know-how and technology that will support next-generation exascale HPC systems. This means operations that are significantly faster compared to contemporary levels, i.e. exaFLOPS or 10^{18} floating-point operations per second. The overall ECOSCALE [269, 171] concept that sets the premise for such an advancement is called UNILOGIC. UNILOGIC represents a high-level description of the key technology behind ECOSCALE's framework and the processing element at the heart of UNILOGIC is the reconfigurable hardware, namely, the FPGA. As explained in more detail in later sections, the ECOSCALE framework's hardware platform consists of a finite number of *Worker* islands, which altogether give rise to a computationally powerful and efficient data mover and provider. Each Worker is primarily comprised of a single FPGA that due to the level of technological advancement associated with these devices, effectively constitutes a System-on-Chip (SoC) in itself. This means that an ECOSCALE Worker includes a software-oriented Processing System (PS) side along with a hardware-oriented Programmable Logic (PL) side. These are combined with local memory to make up a single very powerful processing element.

Consequently, UNILOGIC is the name attributed by the ECOSCALE consortium to the ability to spread-out in the physical domain Worker resources, in one unified manner such that their individual characteristics can be combined in order to execute computationally-intensive tasks very fast. As will be explained later, this is facilitated by UNILOGIC in two main ways. First, data can be transferred close to the physical point where they are to be processed or used remotely, in a transparent manner as if they reside locally. Second, it offers a platform that supports the massively parallel execution of separate tasks for any given application [263] provided that said application does allow for parallelism. With this, idling ECOSCALE resources, i.e., Worker resources that have not been allocated to a particular application, can be assigned to additional applications so that more than one can be harnessed by the ECOSCALE platform at any given time. The implementation of an application's algorithm on the reconfigurable fabric of an FPGA occurs in a dynamic manner using partial reconfiguration.

This chapter discusses the complete ECOSCALE framework which is an expansion on what has been mentioned so far. Specifically, UNILOGIC is a constituent element of the ECOSCALE framework but not its sole element. The overall framework sets up an environment whereby a prospective user can not only execute a computationally-intensive application but can also adapt the algorithmic descriptions of an application so that they are compatible with the ECOSCALE system and execute efficiently. Hence, the overall framework is comprised of a library of reconfigurable cores of popular computationally-intensive algorithms, a runtime system [194] for application execution and the hardware platform that is the physical implementation of the UNILOGIC architecture. Note that in ECOSCALE nomenclature, the reconfigurable cores are known as *accelerators* and that this is how they will be referred to from now on. Furthermore, in the event that the accelerator library is missing the desired algorithm, the ECOSCALE framework offers the ability for the user to introduce their high-level description of said algorithm for automatically generating the low-level reconfigurable hardware implementation of a new accelerator module. A detailed description of the ECOSCALE framework is provided in the following sections.

7.2   The Framework

The eventual aim and resulting product of the ECOSCALE endeavour is the ECOSCALE Framework. This represents all aspects of ECOSCALE and, as a result, includes a software and a hardware side. The software side of the framework comes under the umbrella of our *Runtime System*, and the hardware aspect is represented by the term *Hardware Platform* according to Figure 7.1.

Subsequently, what an ECOSCALE user would experience can be described as follows. The framework accepts a number of candidate applications for execution on the ECOSCALE Hardware Platform. These applications are tackled by the Runtime System, first, by analysing each in terms of its algorithmic description. Note that the ECOSCALE platform has been developed for OpenCL [16] algorithmic descriptions and that this is what is assumed

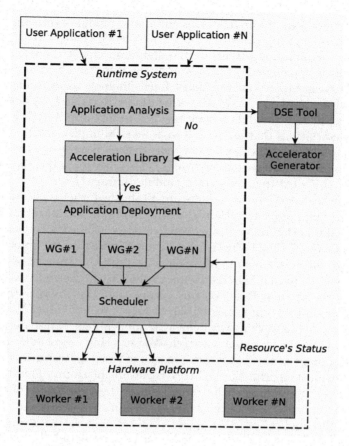

FIGURE 7.1: The ECOSCALE framework.

in this chapter. Nonetheless, the ECOSCALE tools operating under-the-hood are capable of addressing other high-level description languages such as C and C++.

This analysis takes place in order to identify which aspects of the application are suitable for reconfigurable hardware execution and which are more suitable for software execution. For instance, the administrative aspect of an application is commonly not computationally demanding and, therefore, does not require acceleration. On the other hand, any application characteristics that utilise algorithms with computationally-intensive tasks are identified and marked as suitable for Worker execution, i.e., on the Worker's FPGAs.

Subsequently, the framework's acceleration library is queried in order to establish whether the algorithms which have been destined for hardware execution exist in the form of accelerator modules that can be fetched and used further down the line. In the event that the acceleration library does not contain the desired or some of the desired algorithms in the form of equivalent accelerator modules, the framework's automatic module synthesis tool is invoked. This is currently called the *DSE* Tool since it is a tool that performs a design-space exploration process on the high-level OpenCL application algorithms. This is done by automatically applying code modifications that ought to lead to performance optimisations. Each different algorithmic variant is assessed with regard to a range of specification criteria and the optimal is selected for accelerator-module generation. The results from the DSE tool stage are RTL descriptions that are passed to a physical implementation tool [311] that is built around the Xilinx vendor tools to generate the accelerator configuration binaries (called *partial bitstreams*). This bitstream then becomes part of ECOSCALE's existing accelerator modules library.

As soon as all accelerator modules are established, the *Application Deployment* stage comes into effect. Here, the Runtime System takes into account the real-time information related to existing resource utlisation so that it knows which Workers on the Hardware Platform are available for immediate utilisation and which are not. In addition, it gauges the amount of time until resources will be available in the future so that it can schedule tasks to them later on in the process. Based on this information, the Runtime System separates the soon-to-be-implemented algorithms into individual Work-Groups (WGs) since we are working with OpenCL applications. It also decides which Work-Group will be executed on which Worker using a Scheduler module.

The last stage of the framework consists of the accelerator modules hardware deployment on the ECOSCALE hardware platform. This takes place in two stages, first, by partially reconfiguring the logic fabric of the Workers' FPGAs in order to implement the algorithms of interest and, second, by executing the applications themselves. The reconfiguration process is in itself complicated and will be described later in more detail. The ECOSCALE hardware is organised in terms of custom development boards that currently include four Workers, i.e., four FPGAs, called Quad-FPGA Daughter Boards (QFDBs). Subsequently, a collection of QFDBs is what makes up a blade of the ECOSCALE hardware platform and the total number of FPGAs used is going to be 64.

The remaining sections will now provide the reader with a more detailed and insightful account of the major components of the ECOSCALE Framework that were described in this section.

7.3   UNILOGIC and Hardware Platform

The ECOSCALE hardware platform is the physical manifestation of the UNILOGIC architecture, i.e., the technical novelty at the core of the project. UNILOGIC facilitates the seamless sharing of data and FPGA resources across the complete grid of ECOSCALE Workers. Hence, any Worker can access

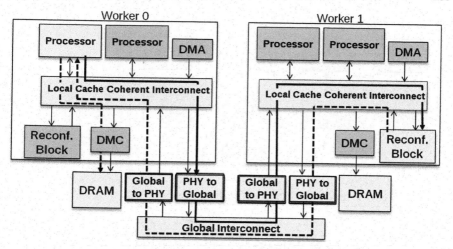

FIGURE 7.2: UNILOGIC concept.

any reconfigurable block, even remote blocks that belong to other Workers. A simplified demonstration of how the inter-Worker communication is realised with UNILOGIC is shown in Figure 7.2. This example involves only two Workers, i.e., *Worker 0* and *Worker 1*, and the aim is for Worker 0 to access the programmable logic, i.e., *Reconfigurable Block*, located at Worker 1 through a physical (PHY) to Global and Global to PHY address translation process. Furthermore, Worker 1 has the option to fetch the data required for processing from the memory (DRAM) of the initiator, i.e., Worker 0, or from its own local DRAM, depending on where they are stored or even any other Worker's memory in a multi-Worker platform. Hence, a complicated and demanding computationally-intensive algorithm can be spread out onto the ECOSCALE hardware resources and executed in-parallel using many Reconfigurable Blocks. In addition, the data required for an algorithm's processing can also be spread out in order to bring them closer to the accelerator modules inside the Reconfigurable Blocks thereby minimising execution times by eliminating time overheads due to lengthy data transfers.

The ECOSCALE platform has a specific definition in the context of the project, i.e., at the lowest level there is the single Worker (represented by the FPGA). The FPGAs used in ECOSCALE are UltraScale+ (US+) devices from the vendor Xilinx. Subsequently, four of these Workers are grouped together on the same board called Quad-FPGA Daughter Board (QFDB). A number of such boards are then placed together in a blade server to complete the hardware platform. The number of combined QFDB boards has been defined as 16 bringing the total number of available FPGAs, i.e., Workers, to 64 for the ECOSCALE hardware platform prototype. Figure 7.3 shows the individual QFDB board with its principal components and, subsequently, how many QFDBs are used to create the ECOSCALE prototype.

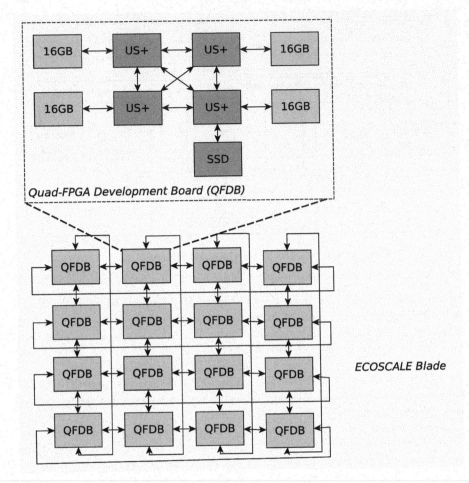

FIGURE 7.3: ECOSCALE prototype.

The QFDB node is a state-of-the-art board that makes use of the capabilities of the FPGAs that it hosts, i.e., it allows for high-computational capabilities as well as high-speed communication between individual Workers. Figure 7.4 shows that the QFDB consists of a custom build board that hosts four FPGAs. The QFDB is a very dense board with all four FPGAs interconnected all to all with two GTH links per connection, i.e., 2×16.3 Gbps, while inter-QFDB communication is realised through one of the four FPGAs, designated as the network FPGA. The network FPGA outputs 10 independent GTH links to SFP+ connectors, i.e., 10×10 Gbps. Moreover, it includes 16 GB of DDR per FPGA in SODIMMs, i.e., 64 GB per QFDB, many power sensors for gathering operational and consumption information, 32 MB of QSPI memory per FPGA, M.2 NVMe storage (SSD) connected to one FPGA, and management Gigabit Ethernet in the Network FPGA.

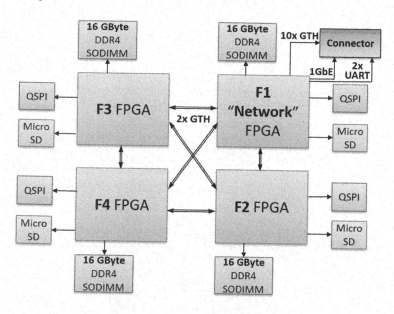

FIGURE 7.4: Schematic of Quad-FPGA development board.

7.4 Runtime System

This section describes the major steps that take place during Runtime System operation. It provides an account of the principal actions necessary for the system to work, and in reality most of those actions will occur automatically and are transparent to the user. Hence, what we provide here is an account of what takes place to an extent under-the-hood so that the reader acquires an informed impression on how the ECOSCALE runtime system works.

Initially, the framework accepts the user applications, as seen in Figure 7.5. Here it is faced with two partitioning problems, first the partitioning of data and second the partitioning of the computations into individual hardware accelerator modules that correspond to individual Work-Groups (WGs) according OpenCL guidelines. With regard to data distribution across the platform storage resources, the mechanism is quite straightforward and has been addressed by the use of a library that allows performing remote memory allocations. Therefore, the job of the runtime is simply to track current memory allocations and executions and then to allocate the available memory according to the allocation algorithm, which takes all of that into account.

With respect to partitioning the computational aspect of an application, the runtime system decides whether the function is to be executed in software or in hardware based on the local status and the status of other Workers in the vicinity. Hence, the runtime identifies which aspects of the application are

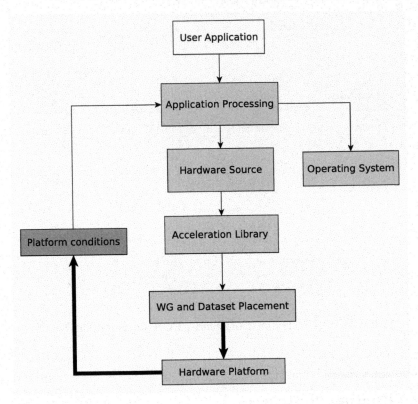

FIGURE 7.5: The major steps of the runtime system flow.

responsible for administrative tasks and, therefore, not as important for hardware implementation, and marks those for execution on the FPGA Operating System (OS). The remaining computationally-intensive tasks are then analysed for hardware implementation through the use of accelerator modules. A useful attribute of the ECOSCALE framework is that the user is provided with the ability to specify functions that can be synthesised in hardware and can be accelerated, on-demand, at runtime, depending on the dynamic execution conditions of the system.

Partitioning computations occurs as follows: when an application has been submitted the runtime uses the information about the previous runs together with information about the current workload (Platform Conditions) in order to make a decision on what sort of hardware accelerators to look for, as well as where to place the computations in the reconfigurable fabric of the available Workers. Note that the placement mechanism is implemented at the NDRange level of granularity. Pending the resolution of a number of problems, the NDRange is decomposed into Work-Groups and the execution is managed at the WG level of granularity. Therefore, the Runtime System is responsible for device discovery, resource management, device reconfiguration, computation submission and execution management.

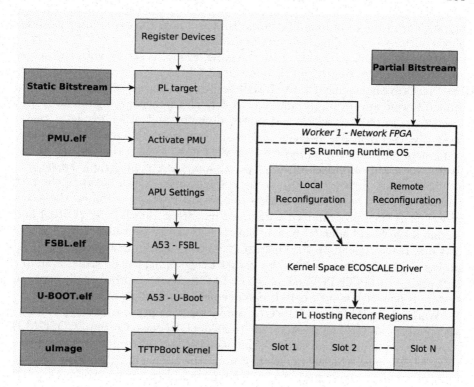

FIGURE 7.6: The Network FPGA flow.

Further expanding the descriptive discussion on the Runtime System, we are now discussing the major runtime steps related to the principal Worker, i.e., Network FPGA, on a single QFDB board, as shown in Figure 7.6. These steps are a detailed account of what occurs the first time that the ECOSCALE hardware platform boots up and, excluding some of the initial steps, what takes place every time the runtime system provides new accelerator modules for local (Network FPGA) or remote execution, on the local QFDB FPGAs or remote QFDB FPGAs.

The ECOSCALE framework uses the Xilinx System Debugger (XSDB) as a tool for the initial system setup. This is a tool that uses a JTAG connection in order to detect all the QFDB board devices that are part of the JTAG chain. This is important since all four Workers, including their PS and PL aspects, are part of that chain. Hence, the first step of the flow is for the individual QFDB devices to register themselves on the JTAG chain. Subsequently, the framework flow selects the PL side of the Network FPGA in order to download and, therefore, implement the reconfigurable architecture of a single ECOSCALE Worker. This is comprised of two parts: i) the static part that includes all hardware peripherals necessary for the Worker to functions, and ii) a dynamic part represented by empty regions called *Slots*, which are the regions in the reconfigurable fabric that will host the accelerator modules through the process of partial reconfiguration.

Subsequently, the PS side of the Network FPGA is set-up with the necessary parameters so that it can sustain the ECOSCALE runtime system. First, the Platform Management Unit (PMU) firmware is downloaded to the FPGA. This firmware is very important since it sets all the necessary parameters related to power states, clock and power domains and other low-level tasks. Recently, this step has become important due to the increased complexity of modern FPGAs such as the US+ used in ECOSCALE that is a complete System-on-Chip comprised of ARM processors and reconfigurable fabric.

Setting the correct ARM processor settings prior to downloading the first and second stage bootloaders is what follows the PMU step. Hence, the ARM processor is then programmed with the First Stage BootLoader (FSBL) and second stage bootloader (U-boot). This brings the system into a state where it is now ready to load up the Operating System's kernel image onto the US+ processor cores. Subsequently, the kernel image is loaded up using a simple network protocol (TFTP) and at the end of the boot up process, the runtime system is now at a state where it can execute all the steps of Figure 7.5.

Worker 1, or the Network FPGA, is now ready to process user applications and perform all the tasks needed for their deployment onto the ECOSCALE hardware. Assuming that the task partitioning has taken place, the runtime then fetches from the acceleration library the accelerator modules (in the form of a partial bitstream) that will be instantiated in the Workers' reconfigurable region slots. Note that in case an accelerator module is missing from the library, the user must create it prior to the execution process using the ECOSCALE DSE and physical implementation tools.

Using Worker 1 as an example, which is shown in Figure 7.6, the runtime decides where an accelerator module will be deployed. It then directs the partial bitstream of interest to the appropriate target reconfigurable slot, which can be local to the Network FPGA or, thanks to the UNILOGIC architecture, anywhere across the ECOSCALE prototype setup. In any case, the runtime OS is equipped with a set of kernel space drivers that can direct the partial bitstream onto the target reconfigurable slot.

7.5 Accelerator Generation

The ECOSCALE framework holds the potential for application execution speed-ups through the implementation of an application's algorithms on reconfigurable hardware. This is made possible by the use of hardware accelerator modules, i.e., low-level partial bitstream representations of computational algorithms commonly found in HPC applications. However, there may be the case, where a desired algorithm does not exist in the ECOSCALE acceleration library. In this case, ECOSCALE offers the ability to automatically generate the low-level representation for an algorithm through its framework, in two distinct steps. First, the algorithm is subjected to a design-space exploration process using ECOSCALE's DSE tool, and, second, the results of the previous step are introduced to an accelerator generator for the actual partial bitstream generation of the accelerator of interest.

7.5.1 DSE Tool

Optimising OpenCL code for FPGA implementation by using Vivado HLS requires significant effort. The task is to guide the compiler to generate optimised compute and memory architectures for each kernel. In other words, the design problem is reversed, from optimising the application for the architecture (on a CPU or GPU) to optimising the architecture for the code (on an FPGA). However, the advantages in terms of energy consumption per kernel execution, and sometimes in terms of performance, more than justify the additional effort. The Xilinx OpenCL high-level synthesis tool, namely Vivado HLS, is used to optimise the code for execution on the Xilinx FPGAs. It requires significant manual annotations of the code with both standard and Xilinx-specific OpenCL attributes in order to achieve a good level of performance. Hence, the DSE tool has been used to enable automated design-space exploration and micro-architecture definition without requiring the designer to specifically know the architectural features of the underlying FPGA. More specifically the ECOSCALE DSE tool consists of a set of programs and scripts that:

- Take as input an FPGA-neutral OpenCL application, composed of a set of kernels and some host code.

- Parse and analyse the OpenCL code to identify areas of the code whose performance, cost or power consumption can be affected by a judicious usage of synthesis directives.

- Generate a full or partial set of design-space exploration options, consisting of legal combinations of those synthesis directives. Of course a partial set ensures faster exploration at the price of some optimality. Currently the designer is in charge of selecting which general areas of optimisation should be explored, e.g., whether loop unrolling should be used at all.

- Execute in a controlled fashion a set of parallel runs of the Vivado HLS synthesis tool, in order to generate a large set of solutions. Parallel execution is essential in order to keep design time limited while exploiting modern multi-core CPUs and multi-CPU compute farms.

- Read back the results, from the synthesis runs in order to i) Compile a set of Pareto-optimal points, i.e., dominating all other micro-architectures in terms of least area or performance, and ii) Make decisions about how to prune the design space in case only partial DSE is performed.

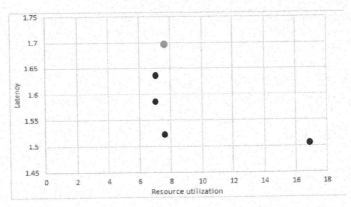

FIGURE 7.7: Design Space Exploration for ECOSCALE oil field simulation application.

Figure 7.7 shows the Pareto space explored by the DSE tool for one of the applications that are used within ECOSCALE to demonstrate the effectiveness of the framework. In this figure, cost is percentage usage of the most constraining resource (among LUT, FF, DSP and BRAM) while latency is total application execution time (including host code and memory transfers). The grey dot shows the result of manual selection of HLS directives by an experienced HLS user.

An interesting aspect is that, by looking at the results of DSE, the tool manages to provide several designs that are better than that achieved by an experienced HLS designer. Nevertheless, a designer familiar with the HLS process can introduce manual code interventions and modifications in order to assist the tool in its effort for design optimisation. This, in fact, has indeed been the case in ECOSCALE, whereby two target application codes were submitted to manual modifications first and, subsequently, were introduced to the DSE process. As a result, the ECOSCALE designers were able to assist the DSE in achieving improved results by changing the code structure, which then helped the tool identify the most suitable directives and the optimal code places in which to introduce them. This shows the effectiveness of design aids in helping the designer identify the most promising areas for application optimisation, rather than fully replacing the designer.

7.5.2 Accelerator Physical Implementation

A typical ECOSCALE Worker includes multiple empty regions called *Slots* acting as resource units to host accelerators at runtime. These slots have the same resource footprints, i.e., the same relative layout of primitives such as look-up tables or memories, and occupy about 50% of the whole chip resources. (See Table 7.1 for the available resources of one slot.) In the ECOSCALE framework, we provide a design methodology to build the accelerators once and relocate them to different slots automatically at runtime as needed.

TABLE 7.1: Available primitives on one resource slot.

Resource primitive	Number	Utilisation
CLB LUTs	32640	11.7%
CLB Registers	65280	11.9%
RAMB36/FIFO	108	12.1%
DSPs	336	13.3%

Moreover, application development can be conducted independently from the system architecture tuning as long as the communication interface between the accelerator and the top-level system remains unchanged. These distinct features are not supported even by the Xilinx state-of-the-art design flow.

After selecting the number of slots which have sufficient resources according to Table 7.1, the RTL code provided by the previous DSE step is first synthesised and then physically implemented. The partial bitstream of the accelerator is then generated and wrapped up with XML metadata for runtime accelerator loading and execution. Reconfiguration is achieved through a process described in the subsequent section. The actual bitstream with accompanying information is specified in the XML file in the proprietary format and, in general, any FPGA can be addressed over UNILOGIC.

7.6 Partial Reconfiguration

The ECOSCALE framework offers the ability to introduce many different computational algorithms on the reconfigurable fabric of the Workers for hardware execution. This is made possible through the use of what is known as partial reconfiguration, i.e., a common and facilitating feature among many contemporary FPGAs. The US+ FPGA that constitutes the basis for the ECOSCALE Worker is manufactured by Xilinx and this vendor has integrated a physical primitive that allows partial reconfiguration in the Programmable Logic of its FPGAs. Hence, the ECOSCALE hardware architecture includes this primitive, which is known as the Internal Configuration Access Port (ICAP). The functionality of ICAP has to do with fetching the partial bitstream and passing it onto the FPGA's PL side, while the information as to where that partial bitstream is to be directed is part of the bitstream itself.

It is worth noting that Xilinx also supports PCAP (Processor Configuration Access Port) quite more actively, which allows partial reconfiguration through the embedded processor. However PCAP cannot be used to support remote partial reconfiguration, which is one of the innovative contributions of the ECOSCALE project, whereas using ICAP we are allowed to so.

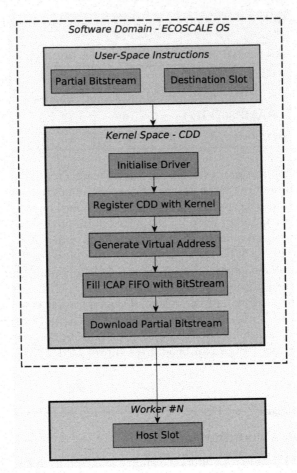

FIGURE 7.8: ICAP flow for partial reconfiguration.

For enabling partial reconfiguration through ICAP, firmware has been developed in the form of a character device driver. This driver allows for a user to instruct the system to download a particular bitstream onto a programmable slot of any ECOSCALE FPGA by operating at the kernel space of ECOSCALE's operating system (OS), as shown in Figure 7.8.

Hence, starting from the software domain, the ECOSCALE framework ensures that a hardware accelerator module partially reconfigures a Worker slot by submitting a set of instructions from the OS User-Space. The user provides the algorithm of interest, and then the runtime system, which is also a User-Space layer, takes over by fetching the partial bitstream that corresponds to the part of the application that is to be implemented in hardware. Subsequently, the two main runtime instructions, shown in Figure 7.8, are submitted to the operating system's kernel space, i.e., the specific partial bitstream of interest as well as the *id* of the destination slot that is to host the accelerator module.

These instructions are passed onto the kernel space and specifically, onto the ECOSCALE Character Device Driver (CDD). As soon as the user-space API addresses the specific driver, a sequence of actions takes place. First, some actions standard to all device-driver operations occur, i.e., the driver is initialised, it is then registered with the OS kernel and, finally, a virtual address is generated so that kernel space operations address hardware modules using virtual and not physical addresses. Next, the Xilinx hardware ICAP module fetches the partial bitstream of interest and stores it on a local FIFO and, subsequently, the ICAP module propagates it onto the reconfigurable fabric of a Worker's Slot of interest. At this point, the reconfigurable logic of the Worker is programmed with the desired accelerator module and is ready for execution during runtime.

7.7   Conclusions

This chapter describes a framework developed in the context of the European Project ECOSCALE. The purpose of this project is to develop a system that can execute computationally-intensive applications at a speed that is acceptable for modern-day HPC standards. Hence, ECOSCALE proposes a novel framework that, based on the capabilities offered by FPGA technology, can massively parallelise the execution of a computationally-intensive application, thereby, laying the foundation for the transition onto the exascale era.

To this end, ECOSCALE, introduces the concept of UNILOGIC, which stands for the ability to harness the potential of a significant pool of processing resources in the form of accelerator modules instantiated in reconfigurable hardware. With UNILOGIC, the system can allocate processing resources according to availability, move data close to said resources so that transfer overheads are eliminated and organise all these actions in an optimal manner and for multiple applications.

The ECOSCALE framework offers a library of popular algorithms ready to be instantiated into the reconfigurable fabric and, furthermore, it offers a set of tools in order to automatically generate new accelerator modules that do not exist in the acceleration library. This way, all designers can use the framework and benefit from the potential that is has to offer.

Acknowledgments

This research project is supported by the European Commission under the H2020 Programme and the ECOSCALE project (grant agreement 671632).

8

A Quantitative Comparison for Image Recognition on Accelerated Heterogeneous Cloud Infrastructures

D. Danopoulos

National Technical University of Athens (NTUA), Athens, Greece

C. Kachris

Institute of Communication and Computer Systems (ICCS), Athens, Greece

D. Soudris

National Technical University of Athens (NTUA) – Institute of Communication and Computer Systems (ICCS), Athens, Greece

CONTENTS

Modern real-world applications in machine learning like visual or speech recognition have become one of the most computationally-intensive applications for a wide variety of fields. Specifically deep learning has gained significant traction due to the high accuracy offered for classification but with the cost of network complexity and compute workload. Recent works have revealed that the domain of deep neural networks is crossing from embedded systems to data centers. Hence, it has been a race between CPU, GPU and FPGA vendors to offer high-performance platforms that are not only fast but also efficient as the energy footprint in the large data centers operating today will be the trade-off for the raw computer power of these platforms. Cloud computing services like Amazon AWS integrate and offer flexibly such modern compute platforms for all kind of tasks, thus facilitating further their development process.

In this chapter we focus on accelerating image recognition on Amazon Compute Cloud, using the Caffe Deep Learning framework [214] and comparing the results in terms of speed and accuracy between different high-end devices (CPU, GPU and FPGA) and network models taking into account the operational cost of each platform.

8.1   Introduction

Emerging applications in robotics, autonomous driving or internet media and data analysis need fast computing systems to perceive the real world. Deep neural networks [335] achieve state-of-the-art perception in both vision and auditory systems. Meanwhile, characterized by their deep structures and large numbers of parameters, deep neural networks challenge the computational performance of today. In this field, Convolutional Neural Networks [242] (CNNs), inspired by the mammalian visual cortex, play a major role because they are highly accurate deep learning networks that represent artificial neurons and can be trained in order to make a desired prediction but with the cost of many calculations. This large computational complexity motivates efforts to accelerate these tasks using hardware-specific optimizations by leveraging different architectures such as CPU, GPU or FPGA. Specifically, FPGAs in the recent years have become popular among computer scientists because of their low power, customizable and programmable fabric for efficiently computing CNNs. Aside from the underlying hardware approaches, a unified software environment is necessary to provide a clean interface to the application layer. This needs to account for several factors, including framework support, different compiler technologies, and optimization support for the underlying hardware engines. The problem becomes more challenging for hosting servers and SaaS providers that deliver exposed Cloud APIs to offload deep learning and inference tasks.

BVLC Caffe is a Deep Learning Framework which has been optimized in CPU and GPU and recently was optimized for FPGA as well, and it is available for Cloud use by many Cloud computing services. So, as the infrastructure, the platforms and the software are provided, end users try to find a trade-off between the operational cost of the Cloud service and the performance of the specific deep learning task.

Thus, based on these facts, this chapter makes the following contributions:

1. We represent a seamless adaptation of the most recent Caffe optimizations for CPU, GPU and FPGA and port the whole framework in Amazon Compute Cloud services.

2. We validate Caffe performance by running benchmark tests on each platform using different network models each time. Also, we make an accuracy measurement of each model using ImageNet validation dataset.

3. Finally, we compare the results from each high-end device taking into account the operational cost of each compute instance and also the accuracy acquired from each model.

8.1.1 Related Work

Most previous work does not include benchmarks of DNNs especially quantitative comparisons, in this case image recognition metrics, on any of the three major platforms used today; CPUs, GPUs, and FPGAs. This work gives a general insight of this application on the latest platforms used today on the Cloud comparing different network models on each platform as well. Thorough steps are described to deploy a DNN using Caffe and run image classification on the Cloud using different Amazon AWS [51] instances. This is an effort which differs this work from others which are mostly typical performance comparison applications, not interesting for the biggest part of the community. Last, as Cloud computing is a fast-emerging field, this work obtains more value.

For example, in [111] the authors compare the performance differences between GPU and FPGA only. Specifically they find that the performance of 6 out of the 15 ported kernels is comparable or superior on FPGA than GPU. Also, the kernels do not include any DNN-related module for image recognition, object detection etc. and the FPGA testing was done on a Xilinx Virtex 7, a smaller FPGA in fabric size than our testing device, the Xilinx VU9P.

Moreover, in [342] the paper describes how to generate accelerators for DNN compute-intensive models using "DNNWEAVER". The measurements include CPUs, GPUs, and FPGAs using Caffe, although they are not deployed on the Cloud or any Cloud computing service but tested "offline". Also, they used the optimized cuDNN library for the Nvidia GPUs, but this is just the basic version while ours include the latest distribution of NVCaffe with device-specific optimizations and the latest versions of libraries (cuDNN 7.3 and CUDA 10).

Furthermore, while there are some papers that make a performance evaluation of DNNs (image training/recognition) such as [341], there is no model comparison. For example, the authors in [341] used the Intel MKLDNN architecture to perform the benchmarks on CPUs only and also do not give results from several network models to compare the differences.

8.1.2 Background

In this section, we first provide a background on deep learning and Caffe framework, the parallelism strategies on software and hardware level and also define the platform heterogeneity that Amazon EC2 Compute Cloud web services provides for acceleration of such applications.

"Just as electricity transformed almost everything 100 years ago, today I actually have a hard time thinking of an industry that I don't think AI will transform in the next several years."

Andrew Ng
Computer scientist and entrepreneur,
co-founded and led Google Brain

Deep Learning. Deep learning is a subset of machine learning, and it has emerged as the most effective method for learning and discerning classification of objects, speech and other types of information resolution. Deep Neural Networks (DNNs), a variant of Convolutional Neural Networks, consist of a large number of artificial neurons distributed across different layers. They are becoming a pervasive tool in a host of application fields aiming to learn from vast amounts of data and classify in near-human accuracy. At the heart of this deep learning revolution are familiar concepts from applied and computational mathematics; notably, in calculus, approximation theory, optimization and linear algebra [197]. These all work together in a digital implementation on software where the purpose many times is obvious but the process of how to achieve it is unclear. Operating like black boxes, neural networks have input layers where users feed their samples and the network performs the necessary calculations to come to the desired output. This operation is called *training* and includes learning from numerous example inputs and modifying a set of weights from a dataset without being specifically programmed but with the process of reverse-composing the gradient of learning given the loss each time [326]. After training, the task is performed on new data through a process called *inference* which is basically the transition of a network from the input stage to the output stage. This *forward pass* computes the "function" represented by the model and typically analyzes-classifies specific data using previously trained DNN models. DNNs trained for image classification have been improved to the point that they are virtually identical to (or even exceed) human accuracy [170, 132].

Deep learning frameworks are sets of software libraries that implement the common training and inference operations. An example is the BVLC Caffe framework which is known for its speed, expressive architecture, extensible code and strong community. It is an open source software, and it is supported across both CPUs and GPUs while recently application developers effectively hid most heterogeneity from the FPGAs and enabled portability across different FPGA devices. Caffe stores, communicates and manipulates all this information as blobs: blob is the standard array and unified memory interface for the framework. The details of blob describe how information is stored and communicated in and across layers and nets. The conventional blob dimensions for batches of image data are number $N \times channel\ K \times height\ H \times width\ W$ which is 4D dimensional and varies according to the type and configuration of the layer. In order to perform an image classification through Caffe framework for example, we need to specify the model definition in prototxt format, the pre-trained model with the weights in caffemodel format, the image mean of the data in binary proto format, the classification categories in a text file and the input image to classify [326]. Caffe provides three APIs for image classification; the C++, Python and Matlab API which has the necessary tools to deploy, test and evaluate DNN models on image classification, multi-label classification, object detection or training/fine-tuning.

Parallelism Strategies. DNN training and inference are two very computationally-intensive tasks and as the network models grow bigger, they consist of many different layers all requiring their own calculations. However, as recent studies showed, most computations usually lie in the convolution layers of a network which are basically a dot product of the features and weights. Many frameworks, including Caffe, convert their convolution operations to large matrix multiplications [105] (see also Figure 8.1) and subsequently make use of common optimized floating-point functions such as Basic Linear Algebra Subprograms (BLAS), GEMM routines, Conv2D operations or even non-linear functions such as Softmax and ReLU. These algorithms perform state-of-the-art performance along with hardware optimizations such as SIMD operations and efficient data transfer mechanisms between GPU or FPGA and CPU.

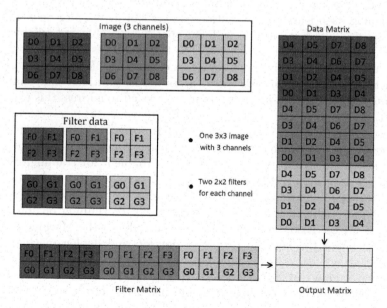

FIGURE 8.1: Caffe's convolution lowering to matrix multiplication [105].

Also, different optimization techniques have been introduced such as 16-bit or 8-bit deep learning operations [177] instead of floating point (32-bit) which are exploited mainly by FPGAs that have re-configurable architecture. They can process more data with less resources, fitting many calculations in single DSPs reducing the energy and computational cost. Especially inference gains leverage from lower bit precision without sacrificing accuracy, and as research has shown, floating point calculations are not required to keep the same level of accuracy. Moreover, training is typically done on a combination of GPUs and CPUs using single-precision floating-point arithmetic (fp32), though the trend is to move towards half-precision floating-point arithmetic (fp16) or less with pruning or quantization of networks to reduce the number of required operations. On top of this, very often frameworks such as MxNet allow scalable parallelization of training tasks using multiple CPUs and GPUs in a distributed manner across large-scale deep architectures.

Heterogeneous Platform Infrastructure. Emerging applications in Cloud computing, IoT and big data analytics have created the need for more powerful data centers that can sustain the huge amounts of the data that are generated by these applications. Therefore, during the last few years, data center operators have started adopting new heterogeneous architectures based on energy-efficient accelerators such as FPGAs and GPUs. Specifically, in deep learning, Cloud-based inference and learning takes advantage of powerful Cloud servers and is by far the most popular design choice when it comes to designing deep learning applications. Users can increase or decrease compute capacity within minutes at will and operate a flexible and reliable Cloud service without the need of ownership of the platforms they use [46]. As a result, unified deep learning configurations combining CPU, GPU and FPGA products are emerging on the Cloud as powerful solutions that enable the compute capabilities and flexibility that are needed to address this expanding market.

Amazon AWS EC2 compute services offer a pre-configured and fully integrated software stack with the latest deep learning frameworks including Caffe. It provides a stable and tested execution environment for training, inference or running as an API service. The stack can be easily integrated into continuous integration and deployment workflows, and it is designed for short and long-running high-performance tasks, optimized for running on Intel CPUs or Nvidia GPUs. Also, recently, Xilinx has introduced the Machine Learning Suite which has the necessary tools to develop and deploy machine learning applications for real-time inference with Caffe on Amazon F1 instances consisting of high-end Xilinx FPGAs. Hence, users can choose the platform of their choice or even combine multiple instances into a single virtual instance for performance-demanding applications depending on the application needs.

8.2   Computational Approaches on Caffe

It is clear that the application and also the project goal are very important in order to choose the right hardware platform. Based on this, platform vendors are building their own advanced high-performance designs on Caffe aiming for fast inference and learning to overcome the continuing exponential increase of the computational load of DNNs.

The latest architecture-specific optimization libraries that also support Caffe framework are known as Intel MKL-DNN (for Intel CPUs), Nvidia cuDNN (for Nvidia GPUs), and Xilinx ML-Suite (for Xilinx FPGAs). These libraries, as recent and personal work showed, have defined the most highly tuned implementations of Caffe that contain standard routines such as forward and backward convolution but are optimized for each specific architecture.

CPU Caffe. The distribution of Intel Caffe [177] improves performance when running the framework in particular Intel Xeon processors. Specifically, in our case, we reproduced the results on Amazon AWS EC2 Cloud specifically on the C5 instance (c5.9xlarge) consisting of 36 Intel Xeon Platinum 8000 vC-PUs with 72 GB total memory. The hardware optimizations included in this distribution have to do with BLAS libraries (switched from Automatically Tuned Linear Algebra System (ATLAS) to Intel MKL-DNN), optimizations in assembly and OpenMP code vectorization. For example, the GEMM function, which is the heart of deep learning and is responsible for most calculations [326, 376], is optimized for vectorization, multithreading and better cache traffic. Also, on the C5 instances with the support of Intel AVX 512 (Advanced Vector Extensions) there is a gain from the better vectorization ratio in the code implementation of Caffe for Intel architecture. Last but not least, this distribution is the first Intel optimized deep learning framework that efficiently supports 8-bit low precision inference [177] and model optimization techniques of convolutional neural networks. The 8-bit models are automatically generated with a calibration process from floating point 32-bit model without the need of finetuning or retraining. Based on the hardware support and Intel Math Kernel Library for Deep Neural Networks (Intel MKL-DNN), Intel Caffe computes the CNN with 8-bit quantization of weights and activations. The distribution also fuses several memory-bound operations of a DNN and folds the learned parameters of batch normalization into convolution kernels to further minimize the execution time of inference. Last, for many DNNs, ReLU is used as a non-linearity placed after the convolution. Intel MKL-DNN supports the merging of convolution and ReLU layers to reduce the memory load and store operations from the default ReLU. It also exploits the particular sparsity properties of specific models such as ResNet type models and applies graph modifications to reduce the total computation and memory accesses [177].

GPU Caffe. NVCaffe [285] is an Nvidia-maintained fork of BVLC Caffe tuned and accelerated for Nvidia GPUs. It includes multi-precision support as well as other Nvidia-enhanced features and offers performance specially tuned for NVIDIA GPU systems. This distribution is composed of specific components such as 16 bit (half) floating point train and inference support, integration with cuDNN and automatic selection of the best cuDNN convolution algorithm, and optimized GPU memory management. Our test platform was the Amazon EC2 P3 instance (p3.2xlarge) which consists of one Tesla V100 GPU on a system with 8 vCPUs and 61 GB Memory.

FPGA Caffe. FPGAs can deliver more flexible architectures, which are a mix of hardware programmable resources, DSPs and BRAM blocks. This flexibility enables the user to reconfigure the datapath easily, even during run time, using partial reconfiguration and lead the desired workload to specific FPGA components. Usually these devices are often used to implement inference accelerators as they have limited memory and can be efficiently configured to minimize the overhead or unnecessary calculations that remain on CPUs or GPUs. However, they are often limited to either integer inference or lower performance floating point compared to GPUs which perform well on single or double-precision floating point mainly in DNN training.

The Xilinx Machine Learning (ML) Suite provides users with the tools to develop and deploy machine learning applications for real-time inference. It also provides support for many common machine learning frameworks such as Caffe, MxNet and Tensorflow as well as Python and RESTful APIs. ML suite has 3 major features that include the xDNN IP which is a high performance general CNN processing engine, the xfDNN middleware which is software library and tools to interface with ML suite and optimize them for real-time inference, as well as the open source support of the distribution. There is also support for the most popular neural networks including AlexNet, GoogLeNet [128] and SqueezeNet [203].

8.3   Performance Evaluation and Quantitative Comparison

In this section, we evaluate the three different approaches on accelerating Caffe as mentioned above (CPU, GPU and FPGA) with the required steps to run on each platform using the Amazon EC2 Compute Cloud instances. Moreover, we evaluate several well-known trained models and compare the performance and accuracy on each model. At the end, after the quantitative comparison, we observe the key differences in each system and we specify the trade-offs regarding performance, model accuracy and device usage cost.

8.3.1 Caffe Metrics on Heterogeneous Cloud

To quantify Caffe's acceleration we measured the net forward and backward performance. Caffe "time" benchmarks model execution layer-by-layer through timing and synchronization. This was very useful to check system performance and measure relative execution times for our models and platforms. In the net forward benchmark, the model passes the information from the input stage to the output in only one direction. This usually happens during the inference (and training as well) when the data is forward propagated in order to compute the output which is the class probabilities. Thus, a single image classification would be the net forward time. Similarly, the net backward is the back propagation of the data which happens many times during training. It is the computation of the gradient given the loss for learning and is very crucial as the network tries to minimize loss and achieves convergence.

On top of this, we try several batch sizes for the image inputs to compare the difference in performance. The batch size defines the number of samples that will be propagated through the network and can have many advantages in inference performance or training accuracy. For example, in inference the device can fill more data in on-chip memory with multiple samples of an input thus reducing the global memory overhead, achieving greater performance per image. Large devices such as GPUs often benefit from these adjustments as they have enough memory to fit many samples for batch processing. Especially, during training, these devices perform well because learning process usually needs big groups of samples. On the other hand, very large batch sizes cause the gradient to be too large and the training cannot converge. But this depends on model and training data because the gradient path can often be good enough. Also, with smaller batch size the model tends to be more noisy and could give good peaks but also bad drops in accuracy. Typically networks train faster with mini-batches because the network parameters are updated after each propagation instead of updating only a few times. Hence, to sum up, if we want to find the performance of classified images per second for example then we can easily solve the equation $\frac{Images}{sec} = (batch\ size) \times (net\ forward\ time)$.

Lastly, we built and ran Intel and Nvidia Caffe on the respective official docker images so as to show how a deep learning model can be built on the spot without worrying about platform specifications and installation setups. Docker can help data scientists build ML/DNN models by installing a relevant docker image without modifying the installation properties depending on the platform thus making it platform independent.

> Docker is a tool designed to make it easier to create, deploy, and run applications by using containers. Containers allow a developer to package up an application with all of the parts it needs, such as libraries and other dependencies, and ship it all out as one package regardless of machine-dependent settings.

8.3.1.1   CPU Performance

Deep learning algorithms are designed to learn and classify quickly. By using clusters of CPUs we were able to run the optimized Intel Caffe distribution and evaluate the results. Our test machine was the Amazon C5 instances, specifically the C5.9xlarge which is made of 36 Intel Xeon Platinum 8000 vCPUs with 72 GB total memory. The hardware optimizations included in this distribution have to do mainly with BLAS libraries (switched from Automatically Tuned Linear Algebra System (ATLAS) to Intel MKL-DNN), optimizations in assembly and OpenMP code vectorization. Also, the cost for these instances was 1.530$ per Hour (at the time of writing this work). However, we compare the CPU performance across three versions of Caffe: the Intel Caffe with MKL-DNN engine, the Intel Caffe with MKL2017 engine and the original BVLC Caffe using OpenBlas and multi-threading.

Build procedure on Intel Caffe is the same as on Caffe's master branch; that is, both Make and CMake compilation methods can be used. However, we built and ran Intel Caffe on docker image for the reasons stated previously. So after installing docker, we downloaded Intel Caffe docker image from docker hub web. We also wanted to share a host data folder for docker container use (models, trained weights, etc.) so we used the "-v" cmd option in the run command like below:

```
$ docker run −v "~/my_models/:/opt/caffe/models"
  −it bvlc/caffe:intel /bin/bash
```

Current implementation uses OpenMP threads. By default the number of OpenMP threads is set to the number of CPU cores. Generally we ran the tests using default configuration but often we had to set each thread to be bound to a single core which is 36 to achieve best performance results because other configurations yielded significant performance decrease. The configurations was set with the OpenMP environment variable "OMP_NUM_THREADS". As for the BVLC Caffe distribution, we manually installed all the Caffe dependencies and libraries and enabled through the Makefile configuration the CPU with OpenBlas support. This enabled us to use the multi-threaded version of Caffe utilizing all 36 cores. We specified the environment variable for OpenBlas "OPENBLAS _NUM_THREADS" to 36 so as to use all threads of our platform and thus have a fair comparison between the three versions.

We picked several popular CNN models for evaluation such as CaffeNet, GoogleNet, SqueezeNet or Resnet50. Moreover, we measured the performance on the C5 instances using three different distributions of Caffe optimized for CPUs. First, the original BVLC Caffe image, then the Intel Caffe with MKL-2017 library and lastly the Intel Caffe with MKL-DNN. The optimization techniques introduced in these benchmarks brought almost up to 1622 $\frac{images}{sec}$ inference performance on CaffeNet using the MKL-DNN library.

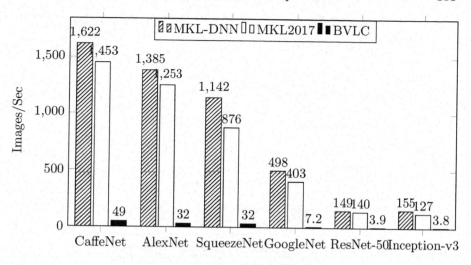

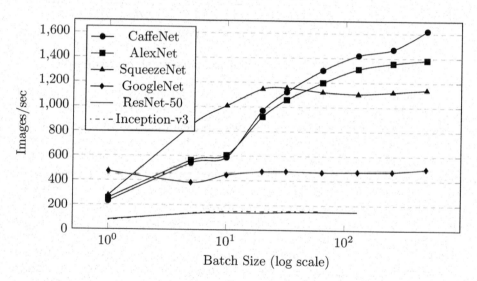

FIGURE 8.2: Top graph: maximum performance achieved in each CPU engine. Bottom graph: MKL-DNN performance per batch size.

Figure 8.2 represents a fair comparison between different model and ditribution performances. In the top figure we can measure the maximum performance achieved with each model running on a different engine every time. All model measurements are derived from 512 batch sizes except for ResNet-50 and Inception-v3 which ran on 128 and 64, respectively. Moreover, in the bottom figure we measure the performance of MKL-DNN engine for different batch sizes. The CaffeNet seemed to be superior for larger batches while in small batch sizes we can clearly notice the lightweight SqueezeNet model advantage.

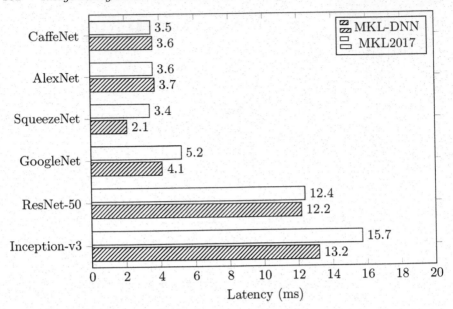

FIGURE 8.3: CPU image recognition latency (shorter is better).

In Figure 8.3 we compare the latency in milliseconds between the different models run on the two engines, MKL-DNN and MKL2017. We deliberately omitted BVLC original Caffe measurements because it was too big to fit in the chart and the small differences on the other engines would not be clear. The point of this measurements is to show the response on each model when a Cloud user demands single classifications and define the best model and engine for this task. This is useful in surveillance systems or generally in critical environments where the maximum performance may be necessary but a fast responsive system would be essentially required. First, it is clear that SqueezeNet has the lowest latency among all network models. Using the MKL-DNN library and SqueezeNet model we achieved only 2.1ms latency. SqueezeNet is a small network designed mainly for mobile devices that have tight computational and memory constraints. As seen in the previous figures, it was successful using small batches but larger batches of images saturated its performance when compared with other networks that increased throughput. Finally, the results show that MKL-DNN compared to MKL2017 is not that superior as in the previous figures where maximum performance is tested. Now, the performance of MKL2017 in latency is competitive where in some cases it is almost the same as MKLDNN or lower.

8.3.1.2 GPU Performance

Modern high-performance computing (HPC) data centers are key to solving some of the world's most important scientific and engineering challenges. NVIDIA accelerated computing platform powers in these modern data centers with the industry-leading applications to accelerate HPC and AI workloads. Specifically, for the GPU metrics, we ran Caffe on Amazon EC2 p3.2xlarge instance which is an 8-vCPU machine with 61 GB total RAM hosting an Nvidia Tesla V100 GPU. This instance supports NVLink for peer-to-peer GPU communication and comes at a cost of 3.06$ per Hour (at the time of writing this work). Each Tesla V100 GPU consists of 5120 CUDA cores and delivers a breakthrough performance of $\sim$ 15 TFLOPS with less power consumption and reduced networking overhead than previous generations. Each GPU-accelerated server can be integrated with several AWS Deep Learning Amazon Machine Images (AMIs) that are pre-installed with popular deep learning frameworks to make it easier to get started with training and inferences such as Caffe, MXNet or Tensorflow.

We ran Nvidia-Caffe, a NVIDIA-maintained fork of BVLC Caffe tuned for NVIDIA GPUs, on the official docker image so as to show how a deep learning application can be built on the spot without worrying about platform specifications and installation setups. We used the . At first, we deployed the NVIDIA Volta Deep Learning AMI which is an optimized environment for running the deep learning framework and HPC containers available from the NVIDIA GPU Cloud (NGC) container registry. The Docker containers available on the NGC container registry are tuned, tested and certified by NVIDIA to take full advantage of NVIDIA GPUs. Then, we acquired the API key from NGC and pulled the latest container image available for the Caffe framework, specifically the version "18.09-py3". Then we simply ran the following command and the environment was ready for running Caffe applications on the Cloud using Nvidia's container image:

```
$ nvidia-docker run --rm --shm-size=1g --ulimit memlock=-1
  -v "~/my_models/:/opt/caffe/models"
  -it nvcr.io/nvidia/caffe:18.09-py3 /bin/bash
```

The benchmark was done under the same rules as on the CPU which is running the Caffe's "time" tool against all the same models and check the performance for different batch sizes. With "nvidia-smi" command we were able to confirm the Nvidia V100 GPU is identified and ready to run. Then, the only difference to Caffe's command was the "-gpu 0" option in order to load the GPU and use the accelerated framework. Also, it is worth mentioning that some models had memory or other issues even at small batch sizes, check failing with error "$axis\_index < num\_axes()$".

Figure 8.4 represents the performance of each model for different batch sizes (where applicable) and the classification latency, at the top and at the bottom of the figure, respectively. As we observe, the performance graph is behaving similarly to the CPU metrics, except for the CaffeNet/AlexNet model which seems to stand out more, achieving a maximum of 13, 054 Img/sec.

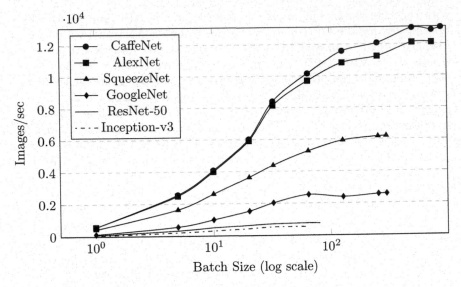

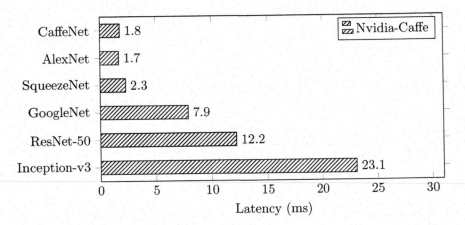

FIGURE 8.4: Top graph: GPU performance per batch size. Bottom graph: GPU image recognition latency (shorter is better).

8.3.1.3 FPGA Performance

Using Cloud computing for deep learning allows large datasets to be easily ingested and managed for training or inference algorithms [203], and it allows deep learning models to scale efficiently and at lower costs using high processing power. By leveraging hardware accelerators, users or developers can develop, test or deploy their deep learning application using FPGAs [128]. One of the advantages of these high-performance platforms is the re-configurable hardware which enables the acceleration of specific computationally-intensive parts of a program that can outperform general-purpose accelerators. For example, complex matrix operations on compute-intensive tasks is an ordinary operation of DNNs, thus FPGAs on the Cloud play a major role in the scalability of these neural networks and also take advantage of the lower bit precision parameters of the models (i.e. 8-bit).

Amazon EC2 F1 is a compute instance with Field-Programmable Gate Arrays (FPGAs) that users can program to create custom hardware accelerators for their applications. F1 instances are easy to program and come with everything in order to develop, simulate, debug, and compile hardware acceleration code, including an FPGA Developer AMI and Hardware Developer Kit (HDK). For the FPGA benchmark of Caffe we used Amazon EC2 f1.2xlarge which consists of a Xilinx Virtex UltraScale+ VU9P FPGA, a 6,800 Digital Signal Processing (DSP) platform. Also, along with this instance there is an 8 vCPU system with Intel Xeon E5-2686 v4 (Broadwell) processors and 122 GB total memory. The instance comes at a cost of 1.65$ per Hour (at the time of writing this work).

For the setup of the DNN benchmark of Caffe we used the Xilinx ML-Suite (xDNNv2) on the "FPGA DEVELOPER" AMI. All necessary steps were followed exactly as described on the github repo from the installation of the dependencies (anaconda, git lfs, etc.) to the environment setup. We successfully ran the example tutorials and where it needed we compiled and quantized our models for inference (8- and 16-bit). First, before being able to deploy networks/models to Xilinx FPGAs it is essential to compile them (offline process, once per model). In this step, the xfDNN Compiler, a high-performance optimizer for Machine Learning inference, converts the model (fusing and merging layers, optimizing memory usage, etc.) so as to increase inference rates and lower inference latency. Next, quantizing 32-bit floating point models to Int16 or Int8 is crucial for the preparation for deployment. The xfDNN Quantizer performs a technique of quantization known as recalibration that can be accomplished in a matter of seconds and still maintain the accuracy of the high precision model. It calculates the dynamic range of the model and produces scaling parameters recorded in a json file, which will be used by the xDNN overlay during execution of the network/model. Lastly, to deploy the model using the xfDNN API we followed the required steps for the correct configuration of the FPGA handle and the execution of the network. To benchmark the models we used the perpetual demo from ML-Suite to test the inference rates for different batches on each model.

Figure 8.5 represents the model performance per batch size (top) and the latency of each model (bottom) using the Xilinx VU9P on Amazon AWS. It is worth mentioning that Xilinx Ml-Suite does not rely on batching (introduces similar performance) because batching does not work exactly as CPU or GPU with weight sharing. FPGA design works at a fine-grained level; each engine runs independently, weight memory is unshared, thus, increasing the number of xDNN engines in the device only increases aggregate device one-batch throughput. Also, a number of metrics above from some models (i.e.,

CaffeNet, AlexNet, etc.) are not evaluated on this work because some models were not fully/partial available (at the time of testing) on the FPGA because specific layers are not supported in hardware. However, on the next paragraph we compare from similar work another FPGA board which have compelling results.

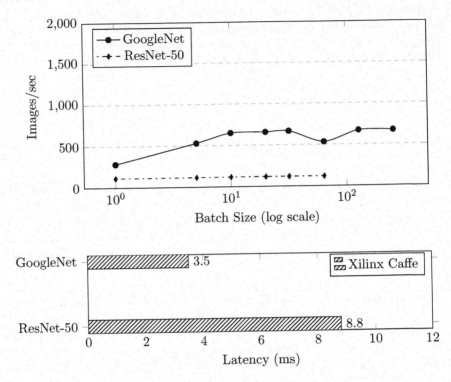

FIGURE 8.5: Top graph: FPGA performance per batch size. Bottom graph: FPGA image recognition latency (shorter is better).

8.3.2   Comparison of Results

After collecting the performance results from the three Amazon AWS platforms we proceeded to measure every model accuracy. Using the ImageNet validation set which consists of 50,000 images, we created the LMDB database in order to use it on Caffe's accuracy measurement tool. We generally used 100 batch size for 500 iterations so as to pass the whole dataset ($100 * 500 = 50,000$). Below, Table 8.1 summarizes the best performance results from every model in each platform along with their respective accuracy and their computational complexity (in GFLOPS).

A lot of useful data and fruitful information can be de derived from the above table. First, we have the model characteristics (GFLOPS, top-5 accu-

TABLE 8.1: Final benchmark results (values in '()' are from other work and FPGA board).

Model Name	Model metrics		Performance (Im/Sec)			Latency (ms)		
	GFLOPS	top-5	CPU	GPU	FPGA	CPU	GPU	FPGA
CaffeNet	0.725	80.3%	1622	13054	N/A	3.5	1.8	N/A
AlexNet	0.728	80.2%	1385	12101	N/A	3.6	1.7	N/A
SqueezeNet	0.388	80.3%	1142	6160	N/A	2.1	2.3	N/A
GoogleNet	1.6	88.9%	498	2557	670 (4127)	4.1	7.9	3.5 (1.18)
ResNet-50	4.1	92.2%	149	765	119	12.2	12.2	8.8
Inception-v3	5.72	94.2%	155	533	N/A	13.2	23.1	N/A

CPU: 36-thread Xeon Platinum @ 1.53$/hour
GPU: Nvidia Tesla V100   @ 3.06$/hour
FPGA: Xilinx VU9P   @ 1.65$/hour *values in '()' taken from [38]

racy) which offers us an insight into the model architecture and efficiency. Then, we can make approximate assumptions on the perfomance each device reaches with the respective model. For example Nvidia Tesla V100 GPU used with CaffeNet, which has a 0.725 GFLOPS model compute workload on single pass, reaches up to 13054 Im/Sec. This means 9464 GFLOPS per second device performance (0.725×13054), not far from the theoretical Single-Precision Performance of 14 TFLOPS.

Also, we can clearly observe that GPU outperforms the other two devices in maximum throughput. While there is an advantage of using GPUs for maximum performance, this is not the case for inference latency. The FPGA is superior to the Image recognition latency achieving very short timings compared with the CPU and GPU alternatives. Unfortunately, as also mentioned earlier many of the FPGA metrics were not available/applicable and as noted in the table, values in '()' are from other work. Specifically the Xilinx white paper [38] makes some metrics on the new Alveo cards, and for a better comparison we included the Alveo U250 benchmarks on GoogleNet which runs on the upcoming xDNN-v3 engine of the ML-Suite.

Figure 8.6 represents the normalized performance speed-up (top) and the normalized performance/cost speed-up (bottom) with baseline of acceleration the CPU values (unit value in y axis). The x axis represent the top-5 accuracy for each available model (their corresponding mark is shown on legend). We can clearly observe that the total throughput is maximized when using the GPU device achieving a $\sim 5.3\times$ speed-up in performance from the CPU baseline. Though, this it not the case when taking into account each device cost where this speed-up is reduced almost in half.

Also, it's worth noticing that we included as reference the Xilinx Alveo U250 FPGA metrics on the GoogleNet model (✱ mark). Although, the design and resource usage on this FPGA is different from our testing board, we can notice the performance advantage of this high end platform.

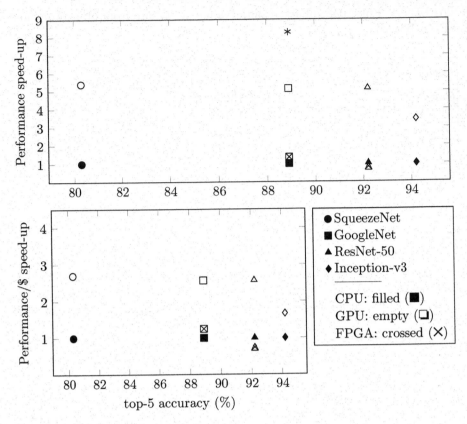

FIGURE 8.6: Top graph: performance speed-up per model. Bottom graph: performance/cost speed-up per model.

8.4   Conclusions

This paper explored the performance of three high-end platforms CPU, GPU and FPGA on the application of image recognition using Caffe framework. The necessary steps were described in order to apply the most advanced device-specific optimization techniques for acceleration. The latest optimized distributions of Caffe for CPUs, GPUs and FPGAs introduced by Intel, Nvidia and Xilinx, respectively, were deployed and tested on the Cloud using Amazon AWS EC2 instances for speed, flexibility and modularity. The tests included image recognition performance between the three platforms under the same conditions using several well-known models for comparison such as GoogleNet, AlexNet or ResNet-50. The aim of this study is to show a developer or an end user can choose the best solution in terms of image recognition speed, latency and accuracy taking into account the device usage cost as well.

The results showed that each of the hardware technologies have advantages and disadvantages depending on the usage scenario. For example the GPU have a high throughput when using with large batch sizes. On the other hand FPGAs are highly customizable devices which have a very short latency and are good for inline processing when compared to the other two devices but are harder to program when application changes functionality. CPUs and GPUs have an easier path to functionality change and are easier to program. It is sure though that deep learning applications are expanding and need these high performance compute platforms. The trend is to use Cloud computing combining all these architectures into a heterogeneous unified configuration where CPUs, GPUs and FPGAs coexist and their uniques compute capabilities are able to address these problems. At application run time the system can choose the best platform whether is DNN training, inference, etc. Thus, flexibility and efficiency are the keys where cooperative platforms with heterogeneous support in the software stack will work together on a full seamless cross-platform environment.

Acknowledgments

This project has received funding from the European Union's Horizon 2020 research and innovation programme under grant agreement No. 687628—VINEYARD, the Hellenic Foundation for Research and Innovation (HFRI) and the General Secretariat for Research and Technology (GSRT), under grant agreement No. 2212-CloudAccel.

9

Machine Learning on Low-Power Low-Cost Platforms: An Application Case Study

A. Scionti, O. Terzo, and C. D'Amico

LINKS Foundation – Leading Innovation & Knowledge for Society, Torino, Italy

B. Montrucchio and R. Ferrero

Politecnico di Torino, Torino, Italy

CONTENTS

The rapid explosion of online Cloud-based services has put more pressure on Cloud service providers for being in condition to satisfy the corresponding huge demand for computing power. While this challenge can be easily tackled by adding more resources, achieving high-power efficiency (i.e., the amount of data processed per watt) is far more complex. In recent years, to cope with the power-efficiency challenge, different hardware systems (e.g., GPUs, FPGAs,

DSPs), each exposing different capabilities and programming models, became part of the standard data center hardware ecosystem. Besides their growing adoption on the core of the Cloud, also edge systems started to embrace heterogeneity to enable data processing closer to the data sources. To this end, lightweight versions of the hardware accelerators available in the data centers, along with number of dedicated ASICs appeared on the market. In this context, performances are not the only metrics to compare different platforms, but energy consumption is also relevant.

This chapter discusses in depth the case study of a machine learning application implemented on a low-power, low-cost platform (i.e., the *Parallella* platform). Computer vision is a widely explored application domain, with large room for optimal implementations on edge-oriented devices, where the limited hardware resources become the main design constraint. First, machine learning approaches are presented, with a focus on Convolutional Neural Networks. Then, different design solutions are analysed, by highlighting the impact on performance and energy consumption.

9.1   Introduction

Pure computing performance are no more the right metric to measure the capabilities of computing systems. The continuous demand for performance has pushed the designers to bump against the power consumption wall. The clock frequency and the number of active cores in modern processors can not increase indefinitely, unless to burn the chips. Traditional multi-core CPU architectures started to show their scalability issues, putting more pressure on the designers to find different solutions. In fact, traditional CPUs are designed to work well on a large number of different applications, limiting the ability of extracting more FLOPS per watt. Conversely, to largely improve (energy) power efficiency, i.e., FLOPS/W, architectural specialisation is required. The way of exploiting specialisation is by coupling multi-core CPUs and accelerators. In this configuration, the CPU drives the execution of the application, requiring the intervention of dedicated accelerators for the most computing demanding parts. Nowadays Cloud computing (CC) is a well-known computing paradigm which is based on the almost infinite availability of resources, that users may quickly, on-demand, acquire and release. With the growing adoption of IoT devices to sense and act on the physical environment, CC progressively enriched its system architecture by integrating such devices, and by pushing computation to be performed on the edge (i.e., closer to the points where data are generated and/or consumed). Among the others, in recent years, applications based on deep neural networks (DNNs—this class includes convolutional neural networks, recurrent neural networks, etc.) received large attention both from industry, Cloud service providers, and scientific community. DNNs are at the basis of a large number of services and applications such as automatic image recognition and classification, recommendation systems, and natural language processing, just to mention few. Looking the edge part of the Cloud, DNNs enable several applications to be efficiently and effectively executed on low power and low cost devices. Among the others, video surveillance and computer vision applications received large attention from industry and academic community. The growing interest for such applications derives from the availability of low-cost devices used to accelerate the most computationally heavy tasks. Such devices range from general-purpose many-core processors, to GPUs and FPGAs, to dedicated ASICs.

Towards the effective exploitation of the hardware accelerators, the exposed programming interface plays a key role. Over the time, GPUs programming interface evolved around two distinct initiatives: *i*) the CUDA programming framework [288], which targets Nvidia GPUs; and *ii*) the OpenCL [223], which aims supporting hardware from different vendors. Both of them extend the set of frameworks (e.g., MPI and OpenMP) that developers can use to parallelize their code. Also, other efforts have been done to effectively support the programming of FPGA devices [52]. Differently from any multi-/many-core device, FPGAs require complex tool-chains to transform HDL codes describing algorithms into the equivalent stream of bits used to configure internal fabric resources on the device). Recent approaches leverages on high-level languages (e.g., C/C++) and transformation tools to provide more flexibility to the programmers. Examples of such efforts can be found in the Intel OpenCL SDK for FPGAs [208], the Xilinx Vivado HLS [334], and the OpenSPL targeting reconfigurable device [64].

Despite the fast progresses showed by compilers in managing complex kernels and transforming them into synthesisable hardware descriptions, general purpose oriented acceleration architectures still receive the largest attention. Conversely, GPUs are one of the most preferred platforms for algorithm acceleration. Being initially designed to process huge amount of floating point operations as found in the computer graphics applications, quickly they have been adopted to process the large number of operations involved in the execution of neural networks. To this end, their internal architecture exposes a very high number of small processing cores, enabling the exploitation of massive parallelism. However, what represents the point of strength of GPUs, also makes complex to efficiently extract the maximum performance. Indeed, GPUs exploit a constrained program execution model, where multiple threads execute in parallel on different portion of the input dataset, making their synchronisation a complex task. General purpose many-cores try to cover the gap between flexible conventional multi-cores and more efficient GPUs, by substituting few complex out-of-order cores with a high number of simple in-order ones. In such case, flexibility is preserved to the programmer, since all conventional parallelization framework are available (e.g., OpenMP, OpenCL, etc.) The *Parallella* single board computer (SBC) provides a 16-cores RISC-like bare metal accelerator within a maximum power consumption of 2W (5W the maximum power consumption of the whole board), i.e., the Adapteva *Epiphany-III*. The capability of the accelerator to execute programs without the overhead of the operating system, make space for achieving high performance in several application contexts.

The aim of this chapter is two-fold: *i*) providing a comprehensive description of state-of-the-art deep-learning algorithms, by focusing on the class of algorithm tailored for image classification (i.e., the convolutional neural networks—CNNs); and describing, as a use case, the implementation of a state-of-the-art CNN architecture on a low power, low cost hardware accelerator. The limited hardware capabilities offered by the selected low power low cost platform (i.e., the Parallella SBC) gives the chance of highlighting, on one hand, the effectiveness reached by CNN models in the image classification context; on the other hand, the platform architectural limiting factors for which improvements may provide large benefits.

9.2   Backgrounds

As suggested in the paper "A brief introduction into Machine Learning" [326] by G. Ratsch, *artificial intelligence* (AI), i.e., the ability of a machine to mimic "intelligent" behaviours of human beings, represents the root of *machine learning* field, which is instead focused on how to make these machines able to "learn". In this regard, the term *learn* refers to an inductive process (*inference*), i.e., the machine learns how to associate given input patterns to a predefined output response. Such ability is built by observing a series of examples regarding a "statistical phenomenon". Thus, machine learning (ML) models refer also to the class of algorithms used to extract the knowledge from the set of input data (i.e., the built model explains the input data). One of the common task for ML algorithms is the classification of input data, i.e., assigning a certain class label according to the (recognised) input pattern. ML algorithms are classified as *supervised* and *unsupervised* methods. In the former case, an initial dataset is used to tune the ML model, by presenting examples of input patterns already correctly classified. In the latter case, the ML algorithm learns how to partition the input dataset into separated classes.

The following are the four most adopted classification algorithms:

- *k-Nearest Neighbour Classification*: such method [116] finds the k points of the training dataset (i.e., features of each sample in the dataset are treated as different dimensions of a multi-dimensional space) that are closer to the analysed input, and assigns a label according to a majority vote (the most present label among the k selected points is used);

- *Linear Discriminant Analysis*: in this method [154], a hyper-plane is computed within the input (dataset) space to minimise the variance between data of the same class (label); while, at the same time, maximising the distance between different classes;

- *Decision Trees*: in this method [321] a tree structure is generated by recursively split the input data space, aiming at creating the purest nodes possible, i.e., nodes only containing points belonging to the same class. Thus, the inference process on a new input point is performed by visiting the built tree model, from the root to the leaves. An extension of such method is represented by the *Random Forest* (RF) algorithm, which during the training phase builds an ensemble of decision trees to improve the accuracy of the classification outcomes.

- *Support Vector Machines*: this approach, also referred to as SVMs [115], originates from the capability of the algorithm to generate a set of hyper-planes in a high (or even in an infinite-dimensional) space, which can be used to classify input data into different classes. Specifically, the name referees to the set of training data (i.e., treated as vectors with regards to a given reference frame) that are strictly required to define the hyper-planes, and that maximise the distance between the two closest vectors of any pair of classes. From this viewpoint, SVMs can be considered as a generalisation of the Linear (Threshold) Machines—LMs;

- *Neural Networks*: these are probably the most-used algorithms to classify input data. Such popularity, also made neural networks (NNs) the main target for hardware acceleration. The method takes inspiration from the organisation of the human brain, trying to re-create it. In a NN, artificial neurons are grouped in multiple layers, each providing a refinement over the final classification, so that the outputs of one layer feed the inputs of the next layer. In such models, the machines learn by assigning specific weights to the connections among the various artificial neurons.

Among the above-mentioned algorithms, neural networks (NNs) received large attention in the last years, so that many variants have been proposed to address different kind of classification problems (e.g., image, voice recognition, etc.), as well as to enrich datasets by generating additional features, starting from a limited number of input data (decoders). As the dataset to classify becomes complex (i.e., high dimensionality, high number of classes), the number of layers grows. Modern NNs use tens of such layers of artificial neurons, leading to the so called *deep*-NNs (DNNs). Conversely, neural networks with very few of such layers (*shallow* networks) provide fast execution, although still able to recognise a very limited number of features in the input data (e.g., the shape of an object—square, triangular, etc.).

Figure 9.1 shows a graphical representation of the artificial neuron (right) along with a general NN architecture (left). Similarly to its biological counterpart, in an artificial neuron, signals arriving at its incoming connections are summed. Aiming at providing different levels of importance to input attributes, input signals' magnitude x_i is weighted. By progressively adjusting the weights w_i used over the whole network, the NN "learns" how to associate

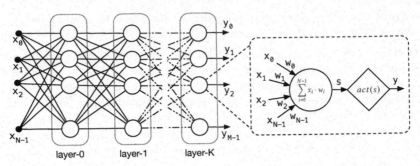

FIGURE 9.1: An example of a neural network architecture (left) along with the representation of the artificial neuron (right).

input patterns to specific output responses. To deal with non-linear effects present in the inputs (e.g., noise), each artificial neuron applies a non-linear function (e.g., ReLU, sigmoid, etc.) to the weighted sum s.

Not only the number of layers makes difference among the NN architectures, but also the way neurons are connected from one layer to the next one. In *Multi-Layer Perceptrons* (MLPs) all the outputs of a given layer are connected to all the inputs in the subsequent layer, thus generating a *fully-connected* architecture. MLPs are difficult to scale up to a large number of layers. To overcome such issue, *Convolutional Neural Networks* (CNNs—ConvNets) use a different approach: each neuron receives only a subset of the outputs from the prior layer, and the subset is generated throughout a convolutional operation. Finally, in *Recurrent Neural Networks* (RNNs), each layer passes to the subsequent one its outputs and a state. By leveraging on the state information, RNNs can reuse the weights across time steps.

9.2.1   Convolutional Neural Networks (CNNs)

Convolutional Neural Networks (CNNs or ConvNets) are ML models in which the information transferring from one layer to the next is defined by a *convolution* operation on a input signal of two independent variables.

Basically, convolution is defined as the integral of the product of a given input signal $x(\tau)$ and a weight signal w, that has been reversed and shifted by a fixed quantity, i.e., $w = w(t - \tau)$. In discrete systems the two functions are sampled, such way the integral operation is substituted by the summation over the variable τ. The operation can be applied for different values of the shifting parameter t, so that an output signal $y(t)$ is generated (see equation 9.1).

$$y(t) = (x * w)(t) = \sum_{\tau=-\infty}^{+\infty} x(\tau)w(t - \tau) \tag{9.1}$$

As mentioned above, the operation can be generalised to the case of an input signal $x(\alpha, \beta)$ of two independent variables. Furthermore, if the weight function is not reflected, the operation becomes known as *cross-correlation*. If the number of samples is finite, then the operation can be conveniently expressed as follows (the operator is also commutative, i.e., $y[\alpha, \beta] = (x \star w)[\alpha, \beta] = (w \star x)[\alpha, \beta]$):

$$y[\alpha, \beta] = \sum_{n=0}^{N-1} \sum_{m=0}^{M-1} x[\alpha + n, \beta + m]w[n, m] \qquad (9.2)$$

When applied to the image processing, the two independent variables are interpreted as the indexes of input image, while the weight function is interpreted as a convolutional filter with a grid arrangement whose indexes are respectively n and m. The application of such an operation to the input image results in an output set of values called *feature map*. The resulting values are also summed with a constant *bias* value δ.

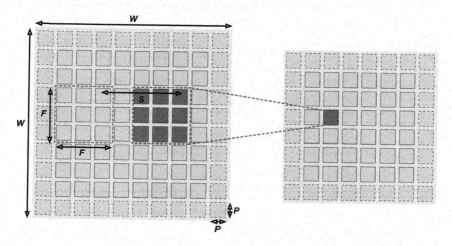

FIGURE 9.2: Local receptive field for the hidden neuron.

In a CNN, artificial neurons of each layer are arranged as a two-dimensional map. Neurons of the first layer capture the magnitude of each cell in the input map (e.g., this corresponds to the pixel intensity of the input image), while the neurons on the other layers progressively capture specific features of the map received from the previous layers. Neurons are connected to those on the next (hidden) layer in such way only a small sub-region of the input two-dimensional map is transferred as input to the next neuron. Such a region takes the name of a *local receptive field*. In a CNN, information transferring is mainly performed by convolving input maps with specified filters. To this end, equation 9.2 represents the core of convolutional layers in CNNs. For instance, for an input map with dimensions of 12×12 and local receptive field of size of 3×3, an hidden neuron could have a connection that looks like the one represented in Figure 9.2.

The number of neurons N_{L+1} in the new layer depends on the number of possible moves that a window can perform on the input area. Such value can be computed by knowing the input size (W), the receptive field size (F), the amount of zero padding applied (P) and the stride length (S). The equation for calculating the number of neurons in the new hidden layer is as follows:

$$N_{L+1} = \frac{W - F + 2P}{S} + 1 \qquad (9.3)$$

Striding concerns the distance between two distinct groups of input cells to which the convolutional filter is applied. As such, stride values largely influence the architecture and performance of the network. Indeed, such parameter controls the way filtering operation are applied to input data. For instance, a stride value $\sigma = n$ implies that convolution and pooling are applied every n input data. For instance, setting $\sigma = 2$ in the convolutional stage allows performing convolution every two input pixels/features. Figure 9.3 depicts the striding process in action.

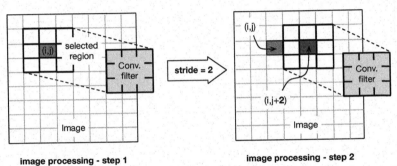

image processing - step 1   image processing - step 2

FIGURE 9.3: Striding mechanism: during processing step-1, the neighbourhood of pixel (i, j) is convolved with a 3×3 filter. In the next processing step, striding of 2 is used, thus the neighbourhood of pixel $(i, j + 2)$ is convolved.

Similarly to input RGB channels that are treated as three overlapped input images, convolutional layers may generate multiple output feature maps which have to be combined. To this end, *pooling* operation is applied to the output feature maps. Pooling operation allows statistically condensation nearby convolved features. Two different types of pooling operations are generally defined:

- *average-pooling*: it computes the arithmetic mean value $p_a[i, j]$ of a subgroup of convolved values, as follows:

$$p_{i,j}^{avg} = \frac{1}{P \cdot Q} \left(\sum_{p=0}^{P-1} \sum_{q=0}^{Q-1} y_{(i+p),(j+q)} \right) \qquad (9.4)$$

- *max-pooling*: it extracts the maximum values $p_m[i,j]$ from a subgroup of convolved values ($\forall p \in \{0, ..., P-1\}$ and $\forall q \in \{0, ..., Q-1\}$), as follows:

$$p_{i,j}^{max} = max\{y_{k,r} : i \le k < i + p, j \le r < j + q\} \tag{9.5}$$

Artificial neurons better mimic the behaviour of brain cells by introducing non-linearity in the way outputs are expressed. To this end, different (many) *activation functions* can be applied. The most common ones are sigmoid, hyperbolic tangent and rectified linear unit (*ReLU*). The latter one offers some advantages over the others: *i*) it is computationally simple, since it involves only a comparison operation; and *ii*) it has a non-saturating shape, which accelerates the convergence of stochastic gradient descent algorithms used during the training phase. Equation 9.6 defines the ReLU function.

$$ReLU(x) = max\{0, x\} \tag{9.6}$$

In addition to convolutional and pooling layers, ConvNets generally implement *fully connected* (FC—all the neurons at a given layer are connected to all neurons in the subsequent layer) and *normalisation* layers at the later processing stages. The purpose of such layers is that of generating the final classification of the input data, as well as to accelerate the network learning by strengthening largely activated artificial neurons. One of the most commonly used normalisation approach uses the *SoftMax* function, that allows extraction of the final output value corresponding to the performed classification. This normalisation function maps the K-dimensional input vector of arbitrary real values $\mathbf{z}$ into a K-dimensional vector of real values in the range $[0,1]$, whose sum equals to 1. The output of such normalisation layer expresses the likelihood that input data presented to the CNN belongs to a given class (see equation 9.7, where $\mathbf{w}$ is the weight vector associated to the input vector $\mathbf{z}$).

$$P(y = j|\mathbf{z}) = \frac{e^{\mathbf{z}^T \mathbf{w}_j}}{\sum_{k=0}^{K-1} e^{\mathbf{z}^T \mathbf{w}_k}} \tag{9.7}$$

9.3 Running CNNs on a Bare-Metal Accelerator: A Case Study

The availability of low-cost digital cameras, along with the growing number applications using such devices (e.g., distributed video surveillance and monitoring applications [338]), reinforced the interest for CNN implementations running on low-power and low-cost computing platforms. In this context, the remainder of this chapter provide a detailed insight of the porting of two state-of-the-art CNN models on the low-power low-cost Parallella SBC platform. A comprehensive analysis of the performance and energy (power) consumption of the system is given, also highlighting the strength and weak points of the implemented solutions. The two selected CNN models are: *i*) the *Tiny Dark-net* trained on the COCO dataset, and *ii*) a custom shallow network trained on the MNIST dataset.

9.3.1   The Parallella Platform

The Parallella SBC platform was born in 2013 following a crowd-funding campaign on Kickstarter launched by Adapteva, aiming to create multi-/many-core processors easy to program, with floating-point support and characterised by a high energy efficiency. The first product announced was the $99 developing Parallella SBC itself. The goal behind this "open source" project was two-fold: democratising the access to high-performance parallel computers, and starting to build a software ecosystem around the Epiphany architecture.

The Parallella SBC is a dual-core system equipped with a dedicated multi-core accelerator (i.e., the Adapteva Epiphany-III—16 cores), which provides bare-metal access to its internal resources. Specifically, the dual-core solution is based on the Xilinx Zynq 7010/7020 System-on-Chip (SoC). It sports two 667 MHz ARM Cortex-A9 cores, which are fully supported by Linux OS, 1 GB of DRAM, 1 Gbps Ethernet, USB and HDMI interfaces. A MicroSD card-reader provides the access to the mass storage through the use of high-capacity MicroSD devices. The Parallella SBC is configured to boot the Linux OS from such a device. Figure 9.4 shows the organisation of the Parallella SBC along with a detailed vision of the architecture of the Epiphany-III fabric.

On the performance front, the 16 RISC-based cores of the Epiphany-III (E16G301) provides up to 32 GFLOPS of raw computing power, with a peak power consumption of less than 2 W. The communication between the host processing system (Zynq) and the accelerator exploits a dedicated interconnect, referred to as the Adapteva e-Link interface. Such interface is implemented as an IP on the Zynq, by using FPGA logic. The dual-core ARM Cortex A9 (each core is a 32-bit out-of-order processor with 32 kB L1 cache per core and 512 kB shared L2 cache) acts as the system (host) processor, while the whole Parallella SBC consumes up to 5 W.

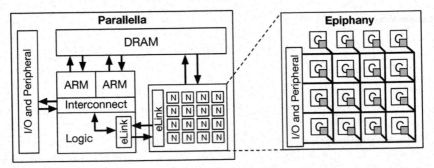

FIGURE 9.4: The Parallella SBC internal organization (left) and the Epiphany accelerator fabric architecture (right).

9.3.1.1 Epiphany-III Bare-Metal Accelerator

The Epiphany-III (hereafter referred to as Epiphany) architecture is formed by a two-dimensional matrix that consists of 16-nodes (namely, *eNodes*) interconnected each other and to the peripheral through a high-speed integrated network-on-chip (NoC), referred to as the *eMesh*. Each eNode contains, in turn, a 32-bit RISC-based processing element, also called *eCore*, a local scratchpad SRAM memory, a direct memory access (DMA) engine, two event timers and a NoC router. Figure 9.5 shows a detailed overview of the Epiphany co-processor.

The eCores are based on a 32-bit superscalar RISC micro-architecture, and are equipped with a 9-port 64-word register files, an integer arithmetic logic unit (ALU) and a floating-point unit (FPU). Worth mentioning is the fact that the instruction set architecture (ISA) is focused mainly on applications largely using floating point operations and C-programmability. All the eCores have also a program sequencer in order to support typical program flows (jumps, branches, etc.) and an interrupt controller. Unlike other micro-architectural designs, the eCores do not use (hierarchy of) cache memories; instead, the integrated scratchpad can be eventually managed by the software to provide such kind of feature. This main lack, is well compensated by the Epiphany support to several program execution models (PXMs), ranging from traditional multi-threading with synchronization barriers and mutexes, to explicit data flow approaches. Furthermore, the internal scratchpad has been designed to maximize the amount of local storage (up to 32 kB per eCore) and memory bandwidth. Indeed, this fast internal storage is split into 4 banks, thus allowing concurrent instruction fetching, data fetching, and multicore communication. This design choice is also reflected in the way address space is exposed to the programmers: the address space is flattened, thus all the eCore may access to the whole 512 kB space (i.e., the summation of all the internal eCore scratchpad space). Also, within this flat address space, each eCore can store both code instructions, data and the memory stack. The address space is further extended by accessing to up to 32 MB (split in two sub-areas of 16 MB each), which are mapped on the external DRAM (host main memory—shared memory). One of the two shared memory partitions is used to store C programming libraries.

To provide high-bandwidth low-latency communication for all the 16 eCores, the Epiphany chip integrates a NoC interconnection, which is actually composed of three separated sub-networks. Each of these sub-networks is based on a 2D-mesh topology, and serves control message exchange, read transactions and write transactions. Through this design choice, packets on the interconnect experience a limited latency of only 1 clock cycle. The addressing on the network is also simplified by using the deterministic X-Y routing protocol, i.e., the eCores specify the position of the destination node in the mesh by using two tags (one for each dimension). The concatenation of such tags provides the unique identifier of an eCore. Least but not last, eNodes ex-

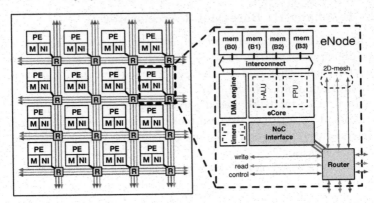

FIGURE 9.5: The internal organization of the Epiphany-III co-processor: on the left side, the 16 eNodes interconnected by dedicated NoCs, on the right side the internal structure of a eNode.

pose two programmable event timers, which can be used to catch performance and specific events during the kernels execution. The list of events that can be monitored is listed in the "Epiphany Architecture Reference" [206] (e.g., number of idle cycles, number of integer instructions issued, traffic on the control mesh, etc.).

9.3.1.2   Application Development

Programmers can use two different approaches to fully exploit Epiphany architectural features: *i*) building the application upon the *Epiphany Software Development Kit* (eSDK), or *ii*) exploiting a higher-level framework, such as as *COPRTHR* and *APL*. The former (eSDK) allows abstraction of specific low-level features (registers operations, interrupts handling, timer setting, mutexes and barriers management, DMA programming) in a convenient way. Conversely, the latter approach provides more abstraction over the underlying hardware, but provides less flexibility in controlling it.

9.3.2   The Tiny-YOLO Convolutional Neural Network

Tiny-YOLO [24] (also referred to as Tiny Darknet) is a small CNN, whose architecture was inspired by SqueezeNet [202] (i.e., a deep-CNN composed by a small number of layers and with a small memory footprint, which came at the cost of reduced accuracy). Similar to SqueezeNet, Tiny-YOLO model only requires 4.0 MB, albeit it maintains a good accuracy level (58.7% and 81.7% of accuracy of prediction, respectively, for the Top-1 and Top-5 classes). Such model was trained against 1000 classes of prediction. The reduced memory footprint makes the Tiny-YOLO model well suited for running on low-power devices with limited amount of memory, such as the case of the Parallella SBC.

TABLE 9.1: Tiny Darknet architecture.

Layers	Filters			Feature Map	
	No.	Size	Stride	Input	Output
conv	16	3 × 3	1	224 × 224 × 3	224 × 224 × 16
max		2 × 2	2	224 × 224 × 16	112 × 112 × 16
conv	32	3 × 3	1	112 × 112 × 16	112 × 112 × 32
max		2 × 2	2	112 × 112 × 32	56 × 56 × 32
conv	16	1 × 1	1	56 × 56 × 32	56 × 56 × 16
conv	128	3 × 3	1	56 × 56 ×16	56 × 56 × 128
conv	16	1 × 1	1	56 × 56 × 128	56 × 56 × 16
conv	128	3 × 3	1	56 × 56 × 16	56 × 56 × 128
max		2 × 2	2	56 × 56 × 128	28 × 28 × 128
conv	32	1 × 1	1	28 × 28 × 128	28 × 28 × 32
conv	256	3 × 3	1	28 × 28 × 32	28 × 28 × 256
conv	32	1 × 1	1	28 × 28 × 256	28 × 28 × 32
conv	256	3 × 3	1	28 × 28 × 32	28 × 28 × 256
max		2 × 2	2	28 × 28 × 256	14 × 14 × 256
conv	64	1 × 1	1	14 × 14 × 256	14 × 14 × 64
conv	512	3 × 3	1	14 × 14 × 64	14 × 14 × 512
conv	64	1 × 1	1	14 × 14 × 512	14 × 14 × 64
conv	512	3 × 3	1	14 × 14 × 64	14 × 14 × 512
conv	128	1 × 1	1	14 × 14 × 512	14 × 14 × 128
conv	1000	1 × 1	1	14 × 14 × 128	14 × 14 × 1000
avg	–	–	–	14 × 14 × 1000	1000
softmax	–	–	–	–	1000

The CNN model accepts, as input, RGB images with size of 224×224 pixels; the network itself presents an architecture composed of 22 layers (see Table 9.1): 16 are convolution layers (CLs) using zero-padding, 4 are max pooling layers (MPLs), one is an average pooling layer (APL), and the last one is a softmax layer (SML). Having such a high number of layers, the number of operations required for an evaluation is, in turn, very high. To keep simple computations, the activation function is always set to *ReLU*.

9.3.3 A Custom Shallow Network

MNIST dataset [13] is one of the most well-known machine learning databases in literature, concerning the recognition of handwritten digits. It has a training set of 60,000 examples and a test set of 10,000 examples, each of them represented by small B/W images. The limited number of classes, as well as the small size of the images represent an interesting basis for studying the development and implementation of a custom shallow network architecture. For the training against such dataset, the open source repository on Github, *darknet_mnist* [4] was used. The network is composed of few layers (see Table 9.2): two convolution layers, two max pooling layers, two fully connected layers and one softmax layer. As for the Tiny-YOLO, also the custom shallow network uses the ReLU as the activation function; although fully connected layers use a *linear* activation function (i.e., *input = output*). Despite the sim-

TABLE 9.2: *darknet_mnist* architecture.

Layers	Filters			Feature Map	
	No.	Size	Stride	Input	Output
conv	32	5 × 5	1	28 × 28 × 3	28 × 28 × 32
max		2 × 2	2	28 × 28 × 32	14 × 14 × 32
conv	64	5 × 5	1	14 × 14 ×32	14 × 14 × 64
max		2 × 2	2	14 × 14 × 64	7 × 7 × 64
full connected	–	–	–	7 × 7 × 64	1024
full connected	–	–	–	1024	10
softmax	–	–	–	–	10

plicity of the network, its main drawback can be found in the large amount of weights required by the two full connected layers, which have a large impact on the overall memory footprint of the network.

9.3.4   Network Training

Training a neural network model is a complex task that generally requires high-performance computing systems to be performed in a reasonable amount of time. For such reason, the Parallella SBC cannot be targeted for training neural networks directly. Such kinds of devices, indeed, are not suitable for this kind of operations, because they are very exorbitant in terms of requested resources. To overcome such limitations, a trend is to train the requested network on higher-performing systems and then to use the "learned parameters" on the embedded devices just for the inference phase. In such a way the less-performing devices remain in charge of the evaluation phase (inference), that is less expensive in terms of both execution time and computational resources.

In our case study, the training of the networks was carried out on a work-station, using the standard library provided by Darknet; then, the results were transferred on the Parallella SBC. In that way two important results were achieved: *i*) the time needed for training was almost solved, using computer systems able to complete such operations in a (relative) short time; *ii*) the parameters used by our implementation are completely compatible with parameters used by the Darknet library.

9.4   Solution Design

The first attempts to implement a working convolutional neural network on the Parallella system just followed a basic and simple idea: *just make it work*, thus not introducing any optimization.

The basic design uses all the 16 eNodes available on the board, following the split of the tasks and the shared memory already described. The approach to convolution is really basic, using a simple direct convolution algorithm, which in turn simply uses four nested `for` loops along the image channels, filters and the two dimensions of the input and output matrices. All the problems related to the memory bottleneck emerged here; indeed, in most of the layers, the operations are completely performed in the shared memory with a tremendous overhead in terms of execution time.

In such solutions, the main program running on the ARM side loads all the parameters in one time in the shared memory, then it sets the co-processor to use all the 16 available eCores. Only two kernels are used to run the entire evaluation process, excluding the APL and the SML that are executed by the ARM cores. On the eCores, all the computations happen in the shared memory: each CL has two local variables referring to the input and output matrices, and which are used to perform all the operations. Unlike the input and output, the convolution parameters (i.e., the filters) are stored in the internal memory. All the layers use a zero-padding of one pixel size. The first padding insertion is executed directly by the ARM cores, during input image loading; conversely, all the subsequent ones are executed directly by the eCores. To this end, a dedicated function copies the input matrices to the secondary memory location (i.e., the one dedicated to the store matrices), adding the additional zero pixels during the memory copy. Specifically, the input value is copied in a memory position that is obtained by shifting by one with regards to the original position.

Thus, the zero-padding operation is reduced to copy the input matrix in a bigger output matrix, shifting its indices by one. In order to be sure that the location of the output matrix will contain only zeros for a perfect zero-padding, a `memset()` operation is performed on that memory location before calling the method just described.

9.4.1 Extended Memory Solution

The basic solution showed how even a low-power system such as the Parallella could run a deep convolutional neural network model. However, at this point the question becomes: how can it made usable, i.e., how performance can be improved in order to be able to use the system in real application contexts?

As shown in the remainder of the chapter, the performance achieved are not in line with those expected for a possible real use, neither using a network like Tiny Darknet, nor using a more suitable smaller network like the custom one trained using the MNIST dataset. The main drawback of the basic design is represented by the need of storing network's parameters and the input and output matrices in the shared memory, which introduces large overheads in terms of processing latency due to the times of reading and writing from the eCores to the external memory and vice versa, even using the DMA engines.

To overcome such a bottleneck, a possible solution is to "extend" the internal memory available for each eCore, so that most of the computations and memory accesses were reverted to the fast internal scratchpads (see fig-

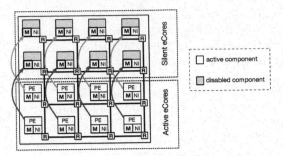

FIGURE 9.6: Graphical representation of the extended memory solution: gray boxes are silent eCores whose memory is bound to one of the active eCores (arrows).

ure 9.6). The implemented solution goes precisely in this direction. Although, reducing the number of active eCores, could impact on the performance (since the degree of parallelism achievable is reduced), a more careful analysis of the situation leads to a positive feedback in switching off some of the eCores and using their internal scratchpads to extend those of the active eCores. Looking at the split of work among the eCores, the reader can understand how the real question was not how to further subdivide the input and output matrices in order to reduce the number of computations to perform, if anything it was mostly how to reach a split that allows keeping almost the whole execution in the internal memory.

Given this premise and keeping in mind communication possibilities offered by the Epiphany co-processor, the formulated hypothesis was: using only half of the available eCores for performing computations and the remaining ones as a storage capacity extension. Obviously, read/write performance between eCores and another one are not the same as the one achieved by directly using their own internal memory; however, they are highly better compared to the use of the shared memory.

9.4.1.1   Implementation

The design choice made led to a new implementation approach. Although the part concerning the main program executed by the ARM side has remained almost unchanged, the implementation concerning the kernels executed by the Epiphany cores has been revolutionized.

The basic idea was to create two groups of eCores:

- *Active group*: such group of cores execute tasks as in the basic solution. What actually changes, is how they perform such tasks, as in the case of convolutional layers, Indeed, each of these cores see a larger internal memory space, giving them the capability of storing more information closer to the processing element;

- *Silent group*: this group of eCores is not involved in the execution of any operation required by the convolutional neural networks. As such, their whole internal memory can be exploited by other eCores to store local variables and executing the required computations directly in such location.

The same number of eCores is assigned to each group, making each active eCore bind with a silent one. This solution allows to doubling the raw internal storage space of the active eCores, thus moving from 32 kB to 64 kB. However, only a portion of this memory space can used to store application parameters, making the whole raw space reaching 48 kB. It is worth to note that silent eCores (not having any kernel to execute) have all their 32 kB available, on the contrary active eCores partially use the internal memory for storing the program code, global variables, etc. From a practice perspective, it was decided to use as active group the first two rows of cores leaving the last two as silent group, More specifically, each active core is bind with its correspondent one on the silent group (i.e., the silent cores are treated, such as the active ones was reflected along the horizontal axis). For instance, starting numbering eCores from left to right and from top to bottom (the first core is numbered as 1), core 1 (top-left on the Epiphany fabric) on the active group is bind with core 1 of the silent group. To make possible such assignment, absolute addresses are used. Thus, a kernel starts with a `switch` construct, which selects the eCores to bind based on the identifier of the core (i.e., the indices of the core on the horizontal and vertical dimension on the NoC) that is executing it. The following code snippet shows such implementation in practice.

As it can be seen, the `switch` block computes the absolute index of the eCore executing the kernel and depending on it `io_sgm_1` and `io_sgm_2` take the absolute addresses of the internal memory of the binding eCore. Worth to note, the silent eCore memory is split in two segments of 16 kB, i.e. `io_sgm_1` and `io_sgm_2` which dynamically store input or output matrices, depending on the executed layer.

Convolutional kernel. Having larger space on more performing memory requires a changing in the way the convolution layer is executed. To this end, input and output matrices are still split into sub-matrices, although they are now stored in the "extended memory".

Also, in this case, the convolution operation considers a direct convolution algorithm, without any specific optimization. However, unlike the case of basic solution, a new step was added, consisting in the copy, using the DMA engines, of the input (sub)matrix from the shared memory to the extended memory location. In that way, all the computations happen directly inside the silent eCore memory, saving the results also inside it. Such results are, at the end, transferred back to the shared memory using the DMA engines. As such, the bottleneck of executing all the operations in the shared memory can be avoid, providing a consistent performance speedup.

Listing 1—Sketch of code for binding Epiphany active eCores with scratchpad memories of silent eCores

```
1    void *io_sgm_1;
2    void *io_sgm_2;
3
4    selection = 4 * e_group_config.core_row +
5                    e_group_config.core_col;
6    switch (selection)
7    {
8      case 0:
9        io_sgm_1 = (void *)0x88800000;
10       io_sgm_2 = (void *)0x88804000;
11       break;
12
13     case 1:
14       io_sgm_1 = (void *)0x88900000;
15       io_sgm_2 = (void *)0x88904000;
16       break;
17
18     case 2:
19       io_sgm_1 = (void *)0x88A00000;
20       io_sgm_2 = (void *)0x88A04000;
21       break;
22
23     ...
24     case 7:
25       io_sgm_1 = (void *)0x8CB00000;
26       io_sgm_2 = (void *)0x8CB04000;
27       break;
28
29     default:
30       break;
31   }
```

This new implementation has another advantage, which is the lack of necessity of a dedicated function for creating the zero padding. As seen with the basic solution, in that case a new function was created in order to add the padding pixels. Although it is effective from a programming viewpoint, the approach is very inefficient since it requires many read/write transactions to happen in the shared memory, as well as requiring the memset() function to directly point to these memory locations. Conversely, with the support of "extended memory" all these transactions are eliminated at the basis. Indeed, the padding pixels can be directly inserted in the input (sub)matrix, once it is copied from the shared memory location to the internal memories of the chip fabric, i.e., shifting the matrix indices of the destination location by one.

Max pooling kernel. The implementation of the max pooling layer is very similar to the one made for the basic solution. The main difference is given by the use of a new step: the input (sub)matrix is copied from the shared memory location to the internal chip memory fabric, by exploiting the "extended memory" approach. As such, almost all the computations are confined within the internal chip memory. The new output (sub)matrix is thus copied back directly in the shared memory.

9.4.2 Optimizing Convolution Operations

Memory extension provides a consistent performance speedup compared to the basic implementation, although it was at the cost of reducing the number of active cores. A further performance improvement can be achieved by enhancing the way convolution algorithm is performed. To this end, the direct convolution appears not very effective, especially on low-power devices where the number of parallel elements is not very high (in fact, parallelization techniques try to unroll loops and assign such independent loop iterations to different cores).

Going in this direction, it was decided to try changing the previous implementations in order to bring them closer to the approach used in the Darknet library. Indeed, the Darknet library does not use a direct algorithm, but rather it uses an optimized version known as the *im2col* method. This method [11] is very popular to speed up the convolution operations, and it is based on the transformation of the dot product between the filter window and the local receptive field into a general matrix multiplication. To this end, all the possible moving windows, i.e., all the regions sampled by the sliding movement of the filter windows, are expanded in memory. The expansion is performed by introducing two new supporting matrices: *i*) the expanded input image matrix, and *ii*) the expanded weights matrix. The latter is simply a row matrix made by serializing all the weights and resulting in a single row vector. Conversely, the former matrix (also referred to as *lowered matrix*) is a matrix having as width the square of the size of a single filter multiplied by the number of input channels, and as height the one calculated as follows:

$$X = \frac{i_h + 2 \cdot p - f_s}{s} + 1$$
$$Y = \frac{i_w + 2 \cdot p - f_s}{s} + 1 \tag{9.8}$$
$$height = X \cdot Y$$

where i_h and i_w are, respectively, the height and width of the input matrix, p is the amount of zero padding to be used in the convolution and s is the stride to use. The matrix product between these two matrices gives the result of the convolution operation, that can be reshaped back by applying the reverse operation (i.e., col2im). The negative aspect of this approach derives from the size of the support matrices used as inputs of the convolutions, which quickly consume large fraction of the memory. For this reason, despite improved performance, it can not be applied to systems with small amount of memory, such as the case of the Parallella SBC.

An alternative method, developed for supporting devices equipped with a small amount of memory is the *memory-efficient convolution* (MEC). Unlike the im2col approach, MEC improves the performance by using support matrices with a smaller size, thus significantly reducing the memory consumption.

9.4.2.1   Memory-Efficient Convolution (MEC)

Memory-efficient convolution [106] is a convolution algorithm aimed at increasing processing speedup while reducing its memory footprint when compared to algorithms such as im2col. Its principle is analogous to that of im2col algorithm, i.e., to transform the dot product between the filter window and the local receptive field into a general matrix-matrix multiplication (GEMM). The difference lies in how the support matrices are created; indeed, MEC uses a different way to lower the input matrix, which allows it to create a much more compact matrix and reduce the memory footprint. The lower impact on memory was the reason behind the decision to use such an algorithm.

Lowered matrix in the MEC algorithm is created starting by splitting the input matrix into sub-matrices with sizes $i_h \times f_s$, where i_h and f_s are, respectively, the height of the input matrix and the size of the convolution filter. Each sub-matrix is generated by sliding the input matrix by s, which is the size of the stride value of the convolution. Once generated, sub-matrices are serialized and copied into one row of the lowered support matrix. Once this operation is completed, MEC further subdivides the newly created lowered support matrix in other partitions, whose size is $o_w \times f_s \times f_s$, and where o_w is the output width and f_s is the size of the filter. Each of these sub-matrices (they are obtained from the intermediate lowered matrix, by shifting it by $s \times f_s$) is multiplied with the support matrix of the weights, which is generated in the same way as im2col algorithm.

The MEC algorithm was implemented only for the custom network trained with the MNIST dataset. Such a decision derives from the sizes of the lowered matrix, which, even smaller when compared to the ones generated by im2col algorithm, still require large memory space. For that reason it was preferred to implement MEC algorithm only in the case of a network where the input matrices are considerably smaller than those used in Tiny Darknet. Compared to other presented implementations, what is really changed is the addition of a new method for generating the lowered matrix, as well as a new method for performing the convolution operation.

Generating the lowered matrix as required by the MEC algorithm is done through a new function referred to as `mec_shaped()`. It uses four nested loops scanning the set of filters, the number of channels and the dimensions of the input matrix. To this end, such function simply performs a copy from one location on the shared memory to the other selected for storing input and output matrices.

The new convolution function, instead, starts from the lowered matrix in the shared memory and copies it into the extended memory of the chip fabric, referring to the partition that must be multiplied by the weights' support matrix. This latter is directly generated in the internal chip fabric memory when copying the weights from the shared memory. After the matrix-matrix multiplication has been performed, the results are saved back in the shared memory.

9.5 Experimental Evaluation

This section offers a view of the experimental results achieved with the proposed solutions. The analysis is divided into two parts: *i*) the analysis of the obtained performance, i.e., the evaluation of the execution time required to run the neural models using the implementations made ad-hoc for the Parallella SBC; *ii*) the analysis of the (energy)power consumed by the Parallella SBC during the execution of such convolutional neural networks.

It was decided to carry out a campaign of 10 measurements for each type of evaluation, reporting the average value along with the standard deviation. What has been tried was to obtain different levels of granularity in the results found. To better express this concept, three types of results have been found with regard to the execution times: *i*) a coarser one which refers purely to the total time required by the entire program to complete; *ii*) a result referring to the time required by Epiphany co-processor to execute, which reports the execution times of every single kernel developed for it and executed by all the involved eCores; finally, *iii*) a finer result that takes into consideration every single operation performed by each Epiphany eCore. To this end, two 32-bit event timers have been used to sample different real-time events within the system, including also a general-purpose clock-cycle counter (see section 9.3.1.1).

9.5.1 Performance Results

First of all, summary charts are shown in Figure 9.7 and in Figure 9.8. These charts report, for both the used neural models, a comparison between the three solutions found (two in the case of Tiny Darknet neural model), concerning the total execution time. From these two graphs a first evidence is how the execution times of the MNIST dataset-based models are much lower than those required for running the Tiny Darknet. That fact can be easily understood since the Tiny Darknet neural model has a much higher number of layers than the one based on the MNIST dataset, in particular the number of convolution layers (the most exorbitant in terms of both resources and required operations) turns out to be much higher. A first logical deduction derived from this consideration is how a system like the Parallella SBC is better suited for running small neural networks, thus having a smaller number of layers to run. A further consideration can be made by observing differences between the

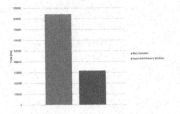

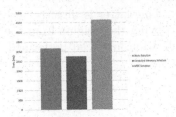

FIGURE 9.7: Tiny Darknet—total execution times comparison.

FIGURE 9.8: MNIST custom model—total execution times comparison.

various implementation of the Tiny Darknet and the ad-hoc MNIST dataset based models. Specifically, the difference in terms of execution times between the basic implementation and the one exploiting the extended memory is much wider with Tiny Darknet with regards to the MNIST-based model. These differences can be seen by comparing Figure 9.9 and Figure 9.10. In the first case it amounts to about 52 s, while in the second it completes in 0.4 s. The reason for this sharp difference can be sought in how the neural models are made. To this purpose, it is worth recalling two fundamental characteristics regarding the networks themselves and the implemented solutions:

- First, there is an important difference in terms of the size of the input and output matrices used by the two models. In the case of Tiny Darknet, the sizes of the intermediate matrices are significantly greater than those required by the MNIST-based model. Indeed, in former case the input matrix is set to $224 \times 224 \times 3$, while in the latter case the size is set to $28 \times 28 \times 3$, that roughly corresponds to a factor of 100;

- Second, these matrices are managed by eCores. In fact, if there is a difference in the management of large matrices between the basic and the extended memory implementations (i.e., in the first case the shared memory is used as storage, while in the second case the scratchpad memory of the silent eCores is used), in the case of matrices of small sizes, these are directly saved in the internal memory of the Epiphany eCore performing the computations.

Starting from these two underlined points, it is understandable that in the case of smaller neural models the Epiphany eCores are able to manage small matrices very often, carrying out the necessary operations on the internal (fast) memory. Even if this were not possible (that is, if the required matrices were not small enough to be stored within eCores), the basic solution would still perform fewer operations in the shared memory by implementing the MNIST-based model, rather than using the Tiny Darknet model. This entails a thinning in terms of execution times between the two solutions. Going in-depth, we can observe the execution times of each implemented layer for both the models, focusing more on the convolution layers, rather than looking the times required to perform each kernel. From this, we can see how in general the main performance differences between the two models

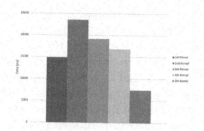

FIGURE 9.9: Tiny Darknet—basic solution—kernel execution times.

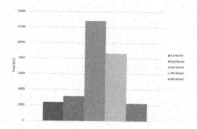

FIGURE 9.10: Tiny Darknet—extended memory solution—kernel execution time.

are in the first executed (kernels)layers. Obviously, it is not possible to state that the execution times are similar for the last performed operations, since these are almost longer than twice as long in the basic solution. However, what can be said is that such enormous initial difference can be ascribed to differences in the two neural network architectures. In these conditions the difference between the two solutions is clear, since in the basic solution the eCores must manage (larger) input matrices (also with a larger number of channels) completely in the external memory, performing a large number of read/write transactions. Conversely, with the extended memory solution, the active eCores are able to perform few transactions on the shared memory thus limiting the overhead. Although the benefits of using smaller network models lead to clear performance advantages, there are still limiting factors that may quickly arise. This generally occurs when the number of transfers from the external memory to that of the dormant eCores quickly increase. This analysis obviously cannot disregard the results achieved with the MEC solution. This solution, as mentioned, has been implemented only for the convolutional network based on the MNIST dataset (see Figure 9.11, Figure 9.12 and Figure 9.13 for a comparison of execution times among the basic, extended memory and MEC-based solutions respectively). Strangely, the results achieved by it in terms of the total time necessary for its execution, are worse than those achieved with the other two presented solutions, although the MEC-based implementation contemplates an optimized version of the convolution operation. As already explained, the optimized convolution algorithms were initially discarded since the performance guaranteed by them are obtained at the expense of greater memory consumption. Therefore, working with a computing system having a small number of resources in this sense, it was preferable to improve performance by following other directions. Nevertheless, after the implementation of the extended memory solution it was decided to proceed with a hybrid solution, integrating the MEC algorithm with an extended chip memory fabric approach (i.e., dormant eCores used as memory extension). To this end, the MNIST-based neural model was targeted, which already requires a smaller amount of memory. Observing the performance data obtained for such implementation, especially those of the individual layers, a strong deterioration in performance can be observed in the execution of the second convolutional layer. That can be attributed in particular to the `mec_shaped()` method used for the generation of the lowered matrix required by the MEC algorithm. The main drawback of this function is not the fact that it works completely in shared memory (indeed, also the method to get the zero padding in the basic solution has the same behavior, albeit it still obtained better results); instead, the issue is in the operations involved, which in turn lead to access non-contiguous memory areas. This mainly depends on the nature of the lowered matrix. Similar behavior manifests itself in executing the core operations of the algorithm, where the lowered matrix is further split for the purposes of being multiplied by the weight matrix. The resulting sub-matrices are constituted in such way non-consecutive accesses in the external SDRAM are performed. This analysis allows that, although such algorithms have been designed to achieve better performance over standard convolution implementations by reducing the overall number of operations, their memory footprint still represents an issue for devices exposing very little memory, thus becoming less effective.

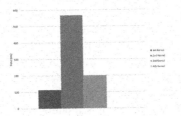

FIGURE 9.11: MNIST custom model—basic solution—kernel execution times.

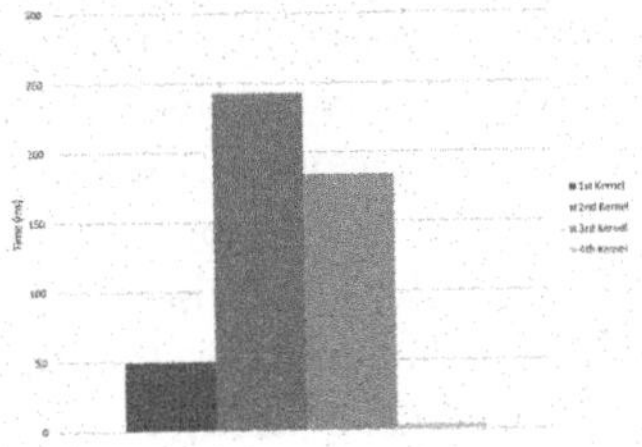

FIGURE 9.12: MNIST custom model—extended memory solution—kernel execution time.

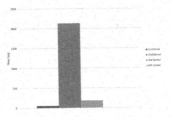

FIGURE 9.13: MNIST custom model—MEC solution—kernel execution times.

Last but not least, an important observation can be made by analyzing the execution times required by the kernels and comparing them with the execution times of the whole application. In such cases, clearly a certain discrepancy emerges. This difference is attributable to the execution times required by the two ARM cores. Although the difference between these results is almost negligible for Tiny Darknet model, it becomes substantial in the case of the MNIST-based model. In such a case, the total execution time becomes almost double if compared to that given by the sum of the times required to run the kernels. This difference is attributable in particular to the fact that the network parameters, in the case of Tiny Darknet model, are loaded at once, at the beginning of the program (i.e., they could be loaded into shared memory only once and then used for recognizing more images), while in the second case, they are loaded in three steps and are interleaved with the execution of the various kernels. This derives from the space itself occupied by these parameters, which makes it impossible to load them into shared memory in a single step. Finally, a general consideration must be given about the execution times of the various layers. In each measurement the standard deviation is generally very low (the maximum value assumed by it is equal to about 5 ms). This is a sign of a good stability of the implemented solutions; at least, as far as performance and execution times are concerned. In general, the reliability of Epiphany architecture can be emphasized.

9.5.2 Power Consumption Analysis

Implementing neural network models on a low-power device, such as the case of the Parallella SBC, demands for an accurate analysis of the overall (power)energy consumption. As initially stated, (energy)power efficiency is becoming ever more a focal point for modern application implementations on heterogeneous systems; as such, convolutional neural network models should be evaluated in this sense. Indeed, despite being often evaluated on the basis of their mere performances, the energy yield must be considered, especially in application contexts such as the industrial one.

As for the performances analysis, also in this case the measurements made refer to the three types of solutions found and to both the used neural models. To measure the consumed power, mainly two approaches were used: *i)* measuring the total power consumption for the entire Parallella SBC, and *ii)* measuring the (energy)power consumption of only the Epiphany chip. In the former case, to measure the total power consumption of the system we fed the board with an external DC power supply keeping the constant voltage at 5.3 V and measuring each time the average electric current absorbed. Then, the power consumption can be traced back to the following equation:

$$power[W] = current[A] \cdot voltage[V] \tag{9.9}$$

In the latter case, we attempted to measure the energy consumption of the Epiphany fabric, and reverted this value to the power consumption by dividing it by the execution time. Since there are no pins on the Parallella SBC dedicated to this purpose, we considered an indirect approach to measure it. The idea followed was inspired by the scientific paper entitled "Instruction level energy model for the Epiphany multi-core processor" [298], and concerns the measure of the energy spent for executing each type of instruction. To this end, instructions are grouped depending on the micro-architectural units involved and the type of events that can be caught by the integrated event-timers. Such characterization has been used to create an energy model for the most important classes of instructions involving, for instance, floating-point operations, integer operations, branches, etc. This model allows to estimate the energy required by an application starting from the instructions' counts. It is worth noting that the event-timers of the Epiphany chip are not able to capture all the possible events related to the execution of all different types of instructions. To overcome such limitation, the original work from Ortiz et al. [298] built specific micro-kernels (i.e., C-based code with in-line assembly to minimize the influence of the compiler), in which the number of instructions to execute can be statically counted. Similarly, the same research work describes an energy model related to remote load and store operations between two eCores (i.e., the model is based on their relative Hamming distance). Combining events measurable with the integrated event-timers and the ones derived by ad-hoc micro-kernels, it is possible to generate a complete energy model. However, in our evaluations, it was not possible to statically count the exact number of instructions per class, but a more rough model can be used, which is based only on timer events and an aggregation of the extracted values.

The high number of performed loops along with the fact we did not use in-line assembly and since relied on the optimizations performed by the compiler, have emphasized the issue. To cope with such limitations we considered the "worst case", thus reporting an upper bound on (energy)power consumed. Specifically for the aggregated instruction, the most wasteful one has been considered to account for the overall energy cost. Event timers used for our purposes are as follows:

- *E_CTIMER_IALU_INST*: counts the integer operations and the load and store operations performed on the local scratchpad. These three operations are all executed by the integer pipeline of Epiphany co-processor (IALU), so the base energy cost taken was that of `local store`.

- *E_CTIMER_FPU_INST*: counts the floating-point operations performed by each eCore.

- *E_CTIMER_E1_STALLS*: counts the number of pipeline stalls.

- *E_CTIMER_EXT_LOAD_STALLS*: was used to calculate the number of external memory accesses while executing load instructions. The value returned by the event timer indicates the number of stalls due to load operations in external memory. Instead, to estimate the actual number of operations, the returned value was divided by the average number of pipeline stalls due to an external data load, which amounts to about 10 (see the Epiphany Architecture Reference [206]).

- *E_CTIMER_IDLE*: counts the number of clock cycles spent in idle.

Event timers do not allow counting of NOP instructions (although they are very few in the generated assembly code), so to count them we opted for a static code analysis. The assembly code was analyzed using a tool supplied by Adapteva: it emulates the modified version of the gcc compiler developed for the Parallella SBC and provides the assembly code as output. From the assembly code it was possible writing (Python) parser to count the number of NOP instructions, as well as branch operations. Concerning the count of write/read operations between eCores, it was possible to use the internal mesh timer. Once set, the timer can monitor several events happening on the internal mesh interconnect, so a parameter (written in the *E_MESH_CONFIG* register) was used to extract only the event count of interest (i.e., accesses to the eMesh for read/write transactions). The energy model related to eCore-to-eCore communication is based on their relative Hamming distance, and the worst case (equal to 3) was used according to the communication patterns defined by the implemented neural models and the basic cost reported by Ortiz et al. [298]. The approach described above allowed to estimate the energy and power consumed by each executed layer.

TABLE 9.3: Tiny Darknet—basic solution—energy consumption.

Operation Type	Avg. Number Of Operations	Base Energy [pJ]	Total Energy [pJ]
IALU Operations	28824604254	47.99	1383292758125.47
FPU Operation	1223578047	29.39	35960958801
Pipeline Stall	32542662920	53.65	1745913865652.63
External Memory Access	29709672950	2054.77	61046544687944.10
Idle Cycle	0	23.59	0
Branch Operation	65214910656.00	154.22	10057443521368.30
NOP	27	17.07	460.89
cMesh Operation	9328558303	419.48	3913143636942.44
	Total Energy [J]		78.182
	Total Execution Time [s]		84.318
	Epiphany Power Consumption [W]		0.927
	Parallella SBC Power Consumption [W]		5.12

TABLE 9.4: Tiny Darknet—extended memory solution—energy consumption.

Operation Type	Avg. Number Of Operations	Base Energy [pJ]	Total Energy [pJ]
IALU Operations	19107873405	47.99	916986844681.96
FPU Operation	1005657407	29.39	29556271192
Pipeline Stall	11072965559	53.65	594064602224.26
External Memory Access	888659711.5	2054.77	1825991315296.12
Idle Cycle	0	23.59	0
Branch Operation	17806218	154.22	9754058186737.92
NOP	21	17.07	358.47
cMesh Operation	2782368566	419.48	1167147966065.68
	Total Energy [J]		14.288
	Total Execution Time [s]		31.578
	Epiphany Power Consumption [W]		0.452
	Parallella SBC Power Consumption [W]		4.95

TABLE 9.5: MNIST custom model—basic solution—Epiphany energy consumption.

Operation Type	Avg. Number Of Operations	Base Energy [pJ]	Total Energy [pJ]
IALU Operations	852938788	47.99	40932532436
FPU Operation	30791963	29.39	904975792.6
Pipeline Stall	325138283	53.65	17443668883
External Memory Access	691576058	2054.77	1421029736984.33
Idle Cycle	0	23.59	0
Branch Operation	1476920448	154.22	227770671490.56
NOP	7	17.07	119.49
cMesh Operation	61124555	419.48	25640528331
	Total Energy [J]		1.514
	Total Execution Time [s]		3.177
	Epiphany Power Consumption [W]		0.476
	Parallella SBC Power Consumption [W]		4.95

TABLE 9.6: MNIST custom model—extended memory solution—Epiphany energy consumption.

Operation Type	Avg. Number Of Operations	Base Energy [pJ]	Total Energy [pJ]
IALU Operations	920768714.5	47.99	44187690609
FPU Operation	30804240	29.39	905336614
Pipeline Stall	479348413.7	53.65	25717042395
External Memory Access	25872053.89	2054.77	53161120171.56
Idle Cycle	0	23.59	0
Branch Operation	2459860992	154.22	379359762186.24
NOP	12	17.07	170.7
cMesh Operation	69045680	419.48	28963281846
	Total Energy [J]		0.532
	Total Execution Time [s]		2.756
	Epiphany Power Consumption [W]		0.193
	Parallella SBC Power Consumption [W]		4.75

TABLE 9.7: MNIST custom model—MEC solution—Epiphany energy consumption.

Operation Type	Avg. Number Of Operations	Base Energy [pJ]	Total Energy [pJ]
ALU Operations	4572171262	47.99	219418498863.38
FPU Operation	30791963	29.39	904975792.6
Pipeline Stall	2847752407	53.65	152781916646.28
External Memory Access	157144781.3	2054.77	322896382189.61
Idle Cycle	0	23.59	0
Branch Operation	2578384384	154.22	397638439700.48
NOP	12	17.07	2014.84
cMesh Operation	444111461	419.48	186295875660.28
	Total Energy [J]		1.280
	Total Execution Time [s]		2.750
	Epiphany Power Consumption [W]		0.275
	Parallella SBC Power Consumption [W]		4.91

From the values reported in Table 9.3, Table 9.4, Table 9.5, Table 9.6, and Table 9.7 the (energy) power consumption is relatively low, approaching the consumption of 1 W only in the case of the basic solution running the

Tiny Darknet model. In the other cases, lower consumption has been found, generally not exceeding 0.5 W. In the best case (i.e., with the extended memory solution running the MNIST-based model) the power consumption fell to 0.2 W, which is far away from the maximum peak consumption (i.e., the peak power consumption is of 2 W, as reported in the datasheet).

A first interesting thing to note is how the solutions optimized for achieving high performance, the extended memory solutions in our case, are also associated to a lower-power consumption. This should not be a surprise considering that, from the energy model previously discussed, the most energy-consuming operation is represented by the access to the external shared memory. From this viewpoint, the basic solution is strongly disadvantaged, since its functional logic performs continuously accesses in shared memory. An interesting point emerges when we look at the MEC-based solution. Although the solution presented not top-in-class performance across the various implementations, its (energy)power consumption is still contained. Again, a limited number of accesses to the external memory positively contributed to the overall energy save.

In addition to the initial premise already made, it is right to spend a few words about the energy model used for Epiphany architecture. As mentioned above the methodology illustrated by Ortiz et al. [298] could not be perfectly applied, due to the complexity of the code developed in comparison with the simple kernels used by the authors. However, the results provided by our approximated energy model are still valid. This is well demonstrated by the empirical measurements made regarding the power consumption of the whole Parallella SBC system. If the instruction-level energy model used was correct, it would mean, from the measurements made, that the power consumption of the remainder of the components would be ~5 W. Besides the application running, in general there are other active processes due to the operating system installed on the Zynq side, as well as other electronic components that may cause such relative high power drawn. It is also interesting to note that ARM cores on the Zynq execute almost the same portion of code (host application); thus keeping the (energy)power consumption of such components (e.g., network interface, Zynq, etc.) almost constant over the time. Moreover, the differences found between different neural models and implementations are in line with the differences found measuring the entire board power consumption. This clearly shows the correlation between the two series of measurements and confirms the validity of the used model.

The model used nevertheless allows a basic idea of how certain approaches are preferable to others with regard to power consumption, based on the type of executed instructions. To summarize, the smaller the number of energy-costly operations are done (e.g., accessing the external memory), the better the performance and (energy)power saving are, although in some case this could lead to a reduction in the number of active cores.

9.6   Conclusion

The main idea behind the work carried out on the Parallella SBC was to perform an evaluation on the use of machine learning techniques on a low-power low-cost system, exploiting as much as possible the resources provided by it. Different approaches have been investigated to optimize the input evaluation time for a single image-recognition operation (i.e., performing the inference), aiming at making the solutions found as usable as possible in a real context: from a basic approach that tries to make the most of the board's multicore architecture to an ad-hoc implementation developed to bypass the main bottlenecks of the device (i.e., the poor amount of internal memory available) trying to exploit the many cores present in a smart way; finally trying to further improve the performances by intervening directly on the algorithm used for the convolution operation. The implementations obtained through these approaches have been tested and evaluated, allowing to make some considerations on their relative strengths and weaknesses. In general, the results obtained were encouraging, especially those related to the use of small neural models, that proved to be the most suitable for the hardware used. Further performance could be obtained moving from a full IEEE-754 single-precision arithmetic to a more efficient weight representation, by using reduced precision floating-point arithmetic or even integer values.

10

Security for Heterogeneous Systems

M. Tsantekidis, M. Hamad, V. Prevelakis, and M. R. Agha

Institute of Computer and Network Engineering – Technical University of Braunschweig, Braunschweig, Germany

CONTENTS

Unlike homogeneous systems, heterogeneous ones, by their very nature, require customization and specialized configuration. These differences may, in turn, require separate verification and certification for each variant, not only increasing deployment costs, but also introducing subtle variations that an intruder may take advantage to attack. For this reason we need common techniques and mechanisms that can ensure the same level of security for heterogeneous systems as we can expect with homogeneous ones. In addition, heterogeneous systems are likely to be more survivable against common mode attacks where a single successful attack strategy can be used to compromise the entire population. Furthermore, heterogeneity may extend beyond a single system, where multiple systems—each one with its own management environment—form a system-of-systems. Different configurations and software drivers will need to be integrated in order to take advantage of the overall system's capabilities. There may be some overlap of tools, however older systems may be running older software for compatibility reasons. This creates the need for general administrators that may have a steep learning curve when trying to familiarize themselves with the super-system.

The remainder of the chapter is organized as follows. In Section 10.1 we present the status of heterogeneous systems today and we provide three well-known examples from a set of diverse domains. In Section 10.2 we detail security-related issues in relation to the examples provided in Section 10.1. Section 10.3 gives a detailed discussion of adaptation within the proposed framework.

Section 10.4 summarizes the research and provides plans for future work.

10.1   Heterogeneous Systems Today

A heterogeneous system architecture integrates heterogeneous processing elements into a single processing environment. It is designed to enable flexible, efficient processing for specific workloads and increased portability of code across platforms. A heterogeneous system comprises of varying types of computing units, interconnects, memory and software entities each with their own architecture.

To demonstrate the multi-platform nature and the effectiveness of our proposed framework (see Section 10.3), we apply it to a diverse set of security-critical, real-world applications. The applications have been chosen from three different domains, namely automotive, buildings and Cloud.

Vehicular systems. Nowadays, a contemporary vehicle contains from 70 to 100 heterogeneous microcontroller-based computers[100], known as electronic control units (ECUs), which are provided from different vendors. Each ECU is hosting one or more software components used to control many functions within the car. However, these applications come from several vendors with various levels of code quality. Safety-relevant functions, such as flight control, anti-lock braking and steering systems, are typically well-engineered and heavily tested, while others, such as the entertainment systems, could be implemented with security and reliability not as prime factors. These ECUs are distributed around the vehicle and interconnected using heterogeneous intra-vehicle networks (i.e. CAN, MOST, FlexRay and Ethernet). In addition, each vehicle can be considered as a part of a heterogeneous system (i.e. Intelligentsia Transportation Systems (ITS)), where the vehicle can communicate with other vehicles, roadside infrastructures, mobile phones, Cloud services, open Internet applications, diagnostics devices, etc.

Smart buildings are examples of heterogeneous systems-of-systems comprising many devices such as surveillance and access control systems, energy management, building automation, smart healthcare systems, etc. These systems are widely distributed and, at the same time, highly interconnected. They are developed using different standards, provide various services, run on different platforms, and operating systems. All these lead to interoperability issues since they are developed in isolation but they need to collaborate with each other. Moreover, the attack surface becomes very wide because all the vulnerabilities of the individual systems can be exploited cumulatively.

The system is as strong as it weakest link, and the attacker can invade a weak sub-system and end up compromising the whole system (stepping stone attack).

Cloud-based systems. A heterogeneous Cloud integrates components by many different vendors, either at different levels (a management tool from one vendor driving a hypervisor from another) or even at the same level (multiple different hypervisors, all driven by the same management tool). Secure execution in such an environment can present many challenges. VM owners use shared hardware and hypervisor infrastructure offered by Cloud service providers, to run their applications on the Cloud. These apps run in isolation, and if the isolation properties are fully enforced then it should be impossible for applications in VMs to modify workloads outside of the allocated resources. However, many different types of attacks have been carried out against Cloud platforms that break free of the VM and target other apps or the underlying layers. This leads to the situation where the management and the security policies of the VM owner need to take into consideration the security-related properties and requirements of the Cloud provider.

10.2 Heterogeneous Systems Issues

New considerations need to be taken into account when developing applications for such heterogeneous systems, compared to homogeneous ones. Different types of computing units in the same system introduce differentiation which may include different instruction set architectures, different memory layout, differences in available libraries/OS services, different interconnects, different performance, etc.

From a security point-of-view, a heterogeneous system is as strong as its weakest sub-component. If an attack compromises a weak component, it may end up infecting the whole system. Each part of the overall system may come from a different source, having its own security measures, communication methods and updating procedures. If a part is provided as a black box, e.g. with proprietary licenses, it may perform actions—possibly malicious—other than the ones originally intended [238, 327].

We have identified four categories that pose security issues in a heterogeneous system, which we deal with under our proposed framework (see Section 10.3):

Security requirements: In a heterogeneous system, components from different vendors come with different levels of security. Various teams with widely varying degrees of competence develop sets of components. Even though these different sets may be securely developed individually, integrating them in the same system could end up leaving the system in a non-secure state. For example, in a vehicular system, an attacker can get benefits from the heterogeneity at the security level of the different ECUs by launching a stepping stone attack on the whole system [189]. Attackers start compromising the weak components or subsystems and use them as an attack surface to plague all related subsystems.

Communication: Parts from different vendors support the use of different communication protocols. However, in a specific setup, these components need to interact with each other transparently in order for the whole system to work properly. This introduces the need for seamless communication during the integration and operation of these components. However, such a process exposes additional security weaknesses, particularly as it must respect all components' requirements. Let us consider a severe weather prediction and response system, for example. Spread across an area, a number of sensors (e.g. wind, light, temperature, humidity, etc.) collect measurements and send them to a Cloud-based processing infrastructure which, in turn, processes the data and controls a number of actuators (e.g. cameras, water blockages, alarm systems, etc.) in real-time, based on the results. The various components of the system may come from different manufacturers, each one supporting different communication protocols (e.g. WiFi, 3G/4G, Bluetooth, etc.) with different vulnerabilities. Each protocol's weaknesses need to be taken into consideration and dealt with, in order to avoid offering a foothold for an attack on the overall system.

Updates: Persistent defense implies that the system software itself and the defensive mechanisms are kept up-to-date with the emerging threats landscape. For instance, all the appliances in a smart home/building have to be secured and always be up-to-date with the latest firmware/patches. This need for regular updates creates an interesting conundrum, namely the need for security updates versus availability. If security updates aren't performed, the devices will remain under threat and malicious actors can leverage their flaws to launch attacks against the overall system [322, 75]. On the other hand, updating a device always carries the risk of ending up with a "bricked" device, i.e. a device that is no longer functional and unlikely to work again. Moreover, updating one component could require updating other components by extension. The failure of updating one component could leave the system nonfunctional. In this case, the roll-back of the updated component to retrieve functionality, could cause several security vulnerabilities.

Management: Another very important feature of a heterogeneous system is the management of its resources. Administrators need to coordinate all the previous aspects of the system (security, communication, updates), so they have a complete picture of its proper functionality and they can interfere whenever something goes wrong.

Furthermore, heterogeneity may extend beyond a single system, where multiple systems—each one with its own management environment—form a super-system. Different configurations and software drivers will need to be integrated in order to take advantage of the overall system's capabilities. There may be some overlap of tools, however older systems may be running older software for compatibility reasons. This creates the need for general administrators that may have a steep learning curve when trying to become familiar with the super-system.

10.3   PDR Secure Framework

Successful defense against cyber attacks relies on **Prevention, Detection** and **Response** strategies that are persistent. We propose a secure framework which analyzes these three strategies within the context of end-to-end security, using the examples from the three domains presented in Section 10.1. We call it *PDR secure framework* (see also Figure 10.1).

Prevention includes all measures that attempt to prevent an attack from succeeding in the first place. In certain cases it is considered better to terminate the task under attack or disable the device rather than allowing it to be taken over. Such a strategy essentially converts a potential compromise into a denial of service attack. Coupled with an effective response mechanism (discussed further on) prevention is the most cost-effective protection mechanism. In high-availability environments such as in electronic trading systems, or vehicular control systems, losing a process or rebooting a system may not be acceptable and as a result special emphasis should be placed on redundancy and system survivability.

The next layer of defense comprises mechanisms that identify an attack against one or more devices. Ideally, **detection** should happen before the attacker has managed to gain a foothold on the victim system. Detecting an attack may include various strategies such as detection of off-nominal behavior (e.g. observing the control flow of a program, or the type of system calls or library calls a process makes), recognition of attack patterns (e.g. signatures used by network intrusion systems, or anti-virus programs).

Once an attack or security breach has been detected, an appropriate **response** should be implemented. This may include actions to limit the damage, i.e. to reduce the scope of the compromise. This may be achieved through the forceful termination of infected or compromised tasks, the isolation at the networking layer (layer 2 or 3) of the compromised host, restarting one or more hosts, and so on. Given the above considerations, a key challenge is the management of all the individual components comprising a heterogeneous system. Developers-administrators need to coordinate all the facets of the system, so

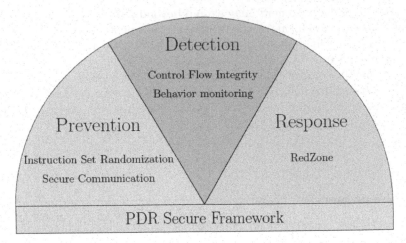

FIGURE 10.1: PDR secure framework.

they have a complete picture of its proper functionality so that they can take the appropriate action whenever something goes wrong.

In the next paragraphs, we detail several mechanisms included in our framework, that help to successfully defend against attacks on a heterogeneous system.

10.3.1 Prevention Mechanisms

When defending a computer system against a potential attack, the initial crucial step is to prevent the attack from happening in the first place. Security measures must be taken to protect the system from any unauthorized use. During this phase, security policies and access control mechanisms must be enforced in order to limit the exposure of the system's resources only to authorized parties. In this section, we present techniques that seek to prevent successful exploitation of a given vulnerability by means of randomization and secure communication.

Instruction Set Randomization (ISR)

ISR [229, 58, 200] is a technique based on diversifying the language the execution environment understands to prevent "foreign" code, which is not expressed in the correct language, from executing. Normally, a CPU architecture provides a runtime that can execute binaries following the CPU's instruction set (x86, ARM, etc.). ISR proposes a runtime that can understand different randomized languages that an attacker cannot possibly know, so he is unable to execute his own code, thus countering *code injection attacks*.

ISR usually relies on simple cryptographic functions to generate random languages with low performance overhead. On binaries, for instance, a random key (*rnd_key*) can be used to randomize every instruction (*instr*), by employing a simple transformation like XOR (*rnd_key* $\oplus$ *instr*), while the reverse process is applied by the runtime after fetching bytes to execute and before decoding them to instructions. The runtime can be implemented directly in hardware [302], which avoids the high runtime overhead for code de-randomization or in software [316] using virtualization. Other implementations use stronger cryptographic algorithms, like AES [57], but incur higher performance overhead and operate on larger code blocks.

Secure communication

Relying only on securing each device within the highly interconnected heterogeneous systems may be not sufficient to ensure the whole system's security. We need to provide compartmentalization in the system network by providing secure communication between the different devices, to prevent security attacks from spreading across the whole system. By using a secure communication we can guarantee the integrity and confidentiality of exchanged data. Moreover, we can ensure the authenticity between the communication ends. Another concern within heterogeneous systems is defining authorization mechanisms, which specify permissions and prohibitions to deny malicious ends from performing unprivileged actions or accessing unauthorized resources.

However, heterogeneous systems have many challenges which make providing such mechanisms a challenging process. As we already mentioned, heterogeneous systems operate using several communication technologies, thus establishing secure communications based on these technologies is a difficult task. Moreover, the computational power of the different nodes within the heterogeneous system could vary. Therefore, not all proposed secure protocols could fit. In addition, in a heterogeneous system a number of security policies coexist and need to be negotiated to reach a common generic security policy for the whole system. Creating this generic policy is difficult, requiring detailed knowledge of possible communication paths between all possible components of the system. Such knowledge is hard to achieve because of the system borders among the different vendors.

Protocols like IPsec are standardized and integrated in many communication protocols to support secure communications that can be adopted in a heterogeneous systems. Using such protocols introduces an overhead which is not negligible. However, we have showed that the overhead imposed by IPsec protocols is small and well within the capabilities of even low cost micro-controllers [191]. To facilitate the definition of a secure communication policy, we have developed a framework to evolve the security policies of intra-vehicle communication [190]. We proposed to build the policy gradually by integrating it through the design and life-cycle of software components. By doing so, the security policy will adapt to changing circumstances during development, integration, and maintenance. Moreover, we will preserve the intentions of the initial designer and ensure the requirements of the actual operational platform. The same procedure can be applied in other heterogeneous systems.

10.3.2   Detection Mechanisms

Prevention mechanisms are not always sufficient to stop all kinds of attacks (e.g. zero day exploits). A second layer of protection is needed. Another category of mechanisms can be adopted to give us a far better understanding of the changes in the state of the various software components within each node of the heterogeneous system, as well as the communication buses between them. In this section we introduce two main mechanisms:

Control-Flow Integrity (CFI)

Software, at the design phase, adheres to the developer's logic. Exploiting an application involves introducing new logic, possibly malicious, and forcing the program to do tasks that it was not originally designed for. One fundamental approach for defending against software exploitation is to enforce a running binary to be contained by the logic stemming from its actual source. This technique is called Control-Flow Integrity (CFI) [36].

CFI thwarts control-hijacking attacks by ensuring that the control flow remains within the control-flow graph (CFG) intended by the programmer. Every instruction that is the target of a legitimate control-flow transfer is assigned a unique identifier (ID), and checks are inserted before control-flow instructions to ensure that only valid targets are allowed. All programs usually contain two types of control- low transfers: *direct* and *indirect*. Direct transfers have a fixed target and they do not require any enforcement checks. However, indirect transfers, like function calls and returns, and indirect jumps, take a dynamic target address as argument. As the target address could be controlled by an attacker due to a vulnerability, CFI checks to ensure that its ID matches the list of known and allowable target IDs of the instruction. An example is shown in Figure 10.2.

CFI, along with W⊕X protection such as Data Execution Prevention (DEP) [48], and ASLR provides strong guarantees regarding the integrity of the protected programs. However, there are two major challenges for the adoption of CFI in its ideal form. First, it requires a complete and precise CFG of the protected application in order to accurately identify all indirect transfer targets and assign IDs. A poor identification of IDs would result in breaking applications. Second, it incurs a non-negligible performance overhead, caused by the introduced checks before indirect control-flow instructions. The larger the number of possible legitimate targets an instruction has, the higher the number of the checks required and the overhead. To overcome these shortcomings, a technique [108] was developed that implements CFI support directly at the hardware level, and modifies all software layers needed for exporting the functionality to running programs. Implementing CFI at

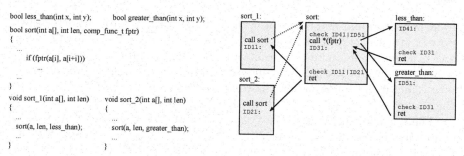

FIGURE 10.2: Example program and its CFG. CFI introduces labels and checks for all indirect transfers. Control-flow transfers checked by CFI are shown in solid lines.

the hardware level has many advantages. First, it is fast compared to existing binary instrumentation techniques, and, second, it is transparent, since all annotation is performed at the compilation phase.

Behavior monitoring

Using only rudimentary detection mechanisms (i.e. based on a static rule) is a risky decision. Behavior monitoring is used mainly to identify the abnormal behavior which could be a signal to new and unknown threats. A task (or a node) may keep complying to its security policy, but it may behave abnormally. For example a task could be authorized to communicate with a certain task on a different node, but it is sending a huge number of requests at a given time. Based on the security policy this activity is normal, but in reality it could be malicious. The task could be compromised and try to mount a Denial of Service (DoS) attack at the remote task. Therefore, behavior monitoring is a suitable mechanism which is used mainly to identify some abnormal behavior that may be a sign of new and unknown threats. This mechanism can be applied in two differed aspects: a) monitoring the behavior of different tasks within each node of the heterogeneous system, b) monitoring the behavior of the communication buses between these nodes/subsystems.

Within each node, many characteristics could be monitored to detect any abnormal behavior of the tasks. For example, some tasks are designed to operate with strict temporal constraints. Thus, any deliberate attacks, such as code injection/reuse attacks where an attacker tries to exploit an existing vulnerability (e.g. buffer overflow) to execute injected or preexisting code to break the system or force it to do malicious actions, will cause temporal misbehavior of these tasks. Therefore, we have proposed a mechanism to monitor the temporal behavior of each task [188]. A security violation is detected as anomalous behavior which can be identified by a breach of the temporal specification of the observed task. Monitoring the power consumption can be used to identify the abnormal behavior of the different tasks in order to detect a variety of attacks [212]. Another approach is to monitor calls to shared libraries and examine their arguments to ensure that they comply with the security policy [267]. An enforcement engine intercepts the call and applies the policy to it. Only if the call is within the restrictions imposed, can it continue with the execution.

The behavior of the network/communication bus is another point which can be monitored in order to identify potential attacks. A Network Intrusion Detection System (NIDS) is a mechanism that inspects the incoming network traffic to detect suspicious payloads before reaching the critical system. The inspection is accomplished by matching the incoming traffic with patterns which are known to be contained in malicious packets. Network packet processing applications can benefit from the support of heterogeneous hardware [371, 301] in terms of exploitation of underutilized computational power, packet throughput and energy consumption.

10.3.3 Response Mechanism

How a system reacts to the detected attack is as critical as the detection of the attack itself. The security policy of the system has to implement different strategies in order to react to an attack and to prevent the failure of the system. The response to a security violation must consider issues such as containment, continued availability, interaction with other subsystems, and, in certain cases, latency [192]. These requirements are in many cases in conflict with each other (for example, security and availability are often in conflict [368]). So stopping (terminating) the software component that triggered the alarm is but one of our options. Moreover, terminating the malicious task without any relevant information about the vulnerability may not be the right choice, as an attacker may repeatedly use the same attack vector to break the newly instantiated tasks. Additionally, in order to better understand the failure and the conditions that triggered it, we need to identify the chain of events that led to the failure as well as the method used by the attacker to exploit the weakness. Regardless of the outcome of our decision, we must make it fast, in order not to violate any potential availability or latency constraints. A further complication arises from the possibility that we are responding to an event that turns out to be a false positive. Unfortunately as the security constraints are tightened, the likelihood of false positives increases. Our strategy to beat this conundrum involves the principle of Red-Zone which we introduced in [319]. Under the Red-Zone regime a task is allowed to breach its designed operational profile until an ultimate limit, but in an observed mode. This window of observation is called the Red-Zone. Whenever the task exceeds the limit, it will be terminated, leaving behind it an audit trail with potential information about the vulnerability or the fault. Figure 10.3 illustrates the possible operational behaviors of a task. The four main areas represent: the intended behavior, the actual behavior, the Red-Zone, and the termination

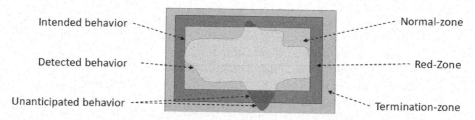

Intended behavior — — — — — — — — — — Normal-zone

Detected behavior — — — — — — — — — — Red-Zone

Unanticipated behavior — — — — — — — — Termination-zone

FIGURE 10.3: Red-Zone principle based on [319].

zone for a given task. The role of the security policy is to define the response
of the system when a task enters the Red-Zone, the conditions under which
a Red-Zoned task may return to normal state, and, finally, the response to
a task breaching its ultimate limit and is terminated by the system. When
a task is terminated, regardless of the cause of termination, security related
or otherwise, the system should respond to the failure by, (i) using an alternate system to provide the functionality that is lost by the demise of the failed
task, or (ii) determining whether the system can continue running without this
functionality (fail-continue), or (iii) forcing the system into a fail-safe condition, i.e. it goes into a mode where the safety of the vehicle and passengers is
assured even if the system can no longer continue to operate.

10.4 Conclusion

In the rush to produce ever more complex systems with ever tighter delivery
deadlines, secure design is often left behind to the detriment of both designers and end-users alike. Fixing security after the design is complete is an
expensive and often impossible task, and we have seen numerous examples
of such failures in the literature and popular press. In this chapter we presented techniques that promote security for heterogeneous systems starting
from prevention, detection and response.

We have shown that monitoring the behavior of a component or task is
achievable without significant overhead, but this relies on effort that needs
to carried out throughout the development process. It is important that key
aspects of what will eventually become the system security policy are defined
as the design evolves by personnel that have intimate knowledge of the design
of each component. In this way security becomes built into the design and not
an unwanted afterthought.

Responding to a security breach is equally important as its discovery in
the first place. An incorrect response may do more damage than the incident
we are responding to. One example is the infamous failure of the maiden
flight of the Ariane V booster which was attributed to the incorrect handling

of floating-point exceptions, which resulted in the shutdown of both primary and secondary Inertial Reference Systems when a counter overflowed, causing the booster to disintegrate in flight [25]. High-value systems (e.g. electronic trading systems, airplanes, etc.) may not be amenable to loss of key subsystems on the mere suspicion of an attack, but would require convincing evidence to the effect that the off-nominal behavior of the component is a threat to the rest of the system. To address these concerns we presented the Red-Zone principle whereby components are marked as suspicious well before they breach the ultimate security envelope forcing the system to terminate them.

The trend towards complex systems with internal networks of heterogeneous computing elements is likely to continue as it allows the designers to chose the most appropriate and cost-effective elements for their designs. However, as we have seen in the previous pages, these systems constitute a very tempting target for potential attackers as they present a complex attack surface. In addition, as these systems, are integrated in expensive platforms (e.g. vehicles), they need to remain capable for a fairly long period of time. As Ross Anderson observed in his speech at Usenix Security 2018, if we replace cars as often as we replace mobile phones, the environmental impact would be enormous. Hence, we need to invest in flexible, dynamic and persistent security mechanisms such as those presented in this chapter.

Acknowledgments

This work is supported by the DFG Research Unit Controlling Concurrent Change (CCC) project, funding number FOR 1800. Additionally, it is supported by the European Commission through the following H2020 projects: THREAT-ARREST under Grant Agreement No. 786890, CIPSEC under Grant Agreement No. 700378, I-BiDaas under Grant Agreement No. 780787, CONCORDIA under Grant Agreement No. 830927 and SmartShip under Grant Agreement No 823916.

11

Real-Time Heterogeneous Platforms

P. Burgio, R. Cavicchioli, and M. Verucchi

Università degli Studi di Modena e Reggio Emilia, Modena, Italy

CONTENTS

While heterogeneous platforms based on many-core accelerators are well established in embedded and high-performance computing domains, they still struggle to find applicability to real-time systems. The main reason is that the architectural complexity of modern accelerators badly fits with traditional design approaches, and analytical frameworks to deliver predictability and timing determinism, the two key properties for real-time architects. Especially, the main issues that prevent a union between the world of real-time systems and the heterogeneous platforms fall into two main macro-areas:

- architectural complexity
- system design and programmability.

Real-time engineers build systems that deliver *a correct result at the correct instant in time*, or at least, within a given time bound. Sometimes, it is preferable that a system not provide an answer at all, if this would mean violating the time constraint. Real-time systems fall into three main categories: *hard-, firm- and soft-real-time*, in decreasing order of criticality and, hence, tightness of requisites. This crucial property is also called determinism or, more often, predictability, and it is typically achieved by design with a thorough analysis phase that involves both hardware and software components of the system, producing the so-called worst-case execution time (WCET) that takes to each component of the system to deliver its result. All of these WCETs are then combined using well-known frameworks to build a complete behavioral model of the whole system. Unfortunately, when hundreds/thousands of cores of the modern accelerators come into play, the complexity of both hardware and software architectures quickly becomes unmanageable for the traditional methods, which in some cases completely fail in modeling the system, making them also inapplicable within industrial settings. Due to that, we cannot rely on them and this means that we need different paradigms and frameworks to WCET analysis. New methodologies and frameworks are being investigated, that potentially can handle such a complexity in a more relaxed manner, based on the fact that, despite the whole system getting tremendously more complex, its components such as computing cores are getting simpler for reasons of energy efficiency, hence they are more analyzable. Research in the field increasingly focuses on the shared resources of the architectures, such as caches, memories and I/O devices, and both academia and industry are well-aware that now it is the memory that is the scarce resource of the system, hence the one that must be adequately managed. In this chapter, the state-of-the-art in the field will be discussed, also comparing to "traditional" methodologies, trying to understand what good can come from them. The second main issue is related to programmability: as shown in previous chapters, the technological shift to heterogeneous platforms requires new programming abstractions and programming models, to help programmers manage the complexity of the system. This is also true in the real-time world, where "traditional" programming models and frameworks must be devised to cope with heterogeneity. Luckily, it seems there is a convergence between what is being done in "non real-time" world, and the needs of the real-time world, so there is already a good basis for research in this area. This chapter also covers this aspect. However, when speaking of programming models for real-time systems, one must always keep in mind that they are tightly connected with the analytical framework to predictability and timing-determinism, and this often results in domain-specific languages and abstractions (e.g., for avionics systems or automotive systems), which are little or no use outside the target application family.

11.0.1 Structure of the Chapter

In the first part, we give concise yet exhaustive basics to understand the concepts of the chapter, even for people that never worked in the real-time domain. We introduce the state-of-the-art frameworks for the analysis, design, implementation and validate industrial-grade real-time systems, focusing on automotive systems as an example. Then, Section 11.1 sums up the most recent research outcomes in the field of system design and WCET analysis. Real-time systems are often safety-critical systems, hence design methodologies and frameworks that prove their effectiveness are hard to overcome and put aside.

However, the rise of heterogeneous, energy-efficient platforms based on many-core accelerators is an attractive choice for building up the next generation of smart real-time devices in the IoT era, hence paving the way for advanced research in the field. The typical example is a vehicle with full or partial self-driving capabilities in a highly-connected environment such as a smart city. Section 11.2 covers these aspects.

Section 11.3 focuses on how programming abstractions are evolving to support programmers. Still sticking to the "autonomous car" example, in the next future hard real-time, safety-critical systems, such as advanced driving assistance systems (ADAS) must co-exist with less critical components, such as infotainment systems. Being able to write efficient code for both of those systems, preserving their properties (high-performance and human-usability in the latter case, safety, and timing-determinism in the former) is an exciting challenge in the field.

We also introduce the most commonly adopted tools and tool flows to design such system look like, and how they will probably look like in the next future. The chapter is inspired by, and focuses on research and industrial activities we carried on within the European Project Hercules, funded by H2020 research and innovation programme under grant agreement No. 688860. Hercules is designed in such a way that all of the technology and methodologies produced can be easily adopted also outside of the specific application domains targeted, that is, automotive and avionics systems.

11.1 Real-Time Task Models

The main goal of real-time systems is not only the correctness of the result, but that it is delivered at the correct instant in time, meaning that once the system receives a stimulus, its response time must be bounded by construction. This is typically achieved by a deep knowledge of both software and hardware system components, which undergo a thorough analysis to understand which is their *worst case response time*, that is, the time to deliver their result. This chapter mainly focuses on the timing analysis and composability of software real-time components, as specified by the H2020 Hercules project.

In traditional approaches for single-core systems, a model of running application(s) is built, and the application itself is split into smaller parts (*tasks*) that are analyzed separately, in isolation, i.e., as running alone in the system with no interference by other components. Single analyses are then composed using well-established mathematical frameworks to derive a global worst-case response time (WCRT, also referred to as worst-case execution time, WCET) of the whole application. They typically assume that an application is split in a set of tasks that either repeat themselves within a given *period* (periodic tasks), or that might occur sporadically, e.g., to react to external stimuli, such as a hardware interrupt. Those latter are called sporadic tasks. Either periodic and sporadic tasks must deliver a result within a given amount of time, called deadline. Missing one deadline represents a failure in real-time systems, which are classified as follows:

- Hard Real-Time — missing a deadline is a total system failure and can lead to catastrophic events, e.g., putting human life in danger

- Firm Real-Time — infrequent deadline misses are tolerable but may degrade the system's quality of service. The usefulness of a result is zero after its deadline.

- Soft Real-Time — the usefulness of a result degrades after its deadline, thereby degrading the system's quality of service.

The main goal of the real-time analysis is to schedule all the running tasks on the available cores in the system so that the timing requirements of the application are satisfied. In the context of multitasking systems, the scheduling policy is normally priority driven (pre-emptive schedulers), so that each task can be, removed the assigned computing unit (pre-empted) if a higher priority task requires it. In real industrial settings, it is possible that a mixture of hard real-time and non-real-time applications co-exist in the same platform, and that hard real-time applications are at different levels of criticality. These are often referred to as mixed-criticality systems.

We now describe the most commonly acknowledge task/system models in the field, and then introduce how traditional WCET analysis looks like.

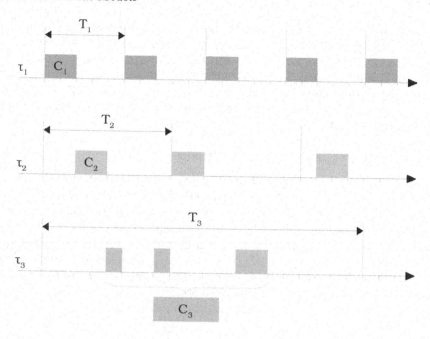

FIGURE 11.1: Three Liu and Layland tasks scheduled on a single processor.

11.1.1 Job-Based Models

11.1.1.1 Liu and Layland (Periodic, Aperiodic, Sporadic)

In 1973 Liu and Layland [254] introduced the first periodic task model. *Periodic tasks* consist of an infinite sequence of identical activities, denominated instances or jobs, that are regularly activated at a constant rate. Each periodic task $\tau_i = (C_i, T_i)$ is characterized by its period T_i and its *worst-case execution time* (WCET) C_i ($C_i \leq T_i$). Deadlines are implicit. Each task generates an infinite sequence of jobs $J_1, J_2, ...$, each job $J_i = (r_i, c_i, d_i)$ with release time r_i, execution time c_i ($c_i \leq C_i$) and deadline $d_i = r_{i+1} = r_i + T_i$. A generalization to this model is with explicit deadlines. In this case each task is defined by a triple tuple $\tau_i = (C_i, D_i, T_i)$ and for each job d_i becomes $d_i = r_i + D_i$. *Aperiodic tasks* consist of an infinite sequence of identical jobs, but their activations are not regularly interleaved. In the *sporadic tasks model*, introduced by Mok [277], sporadic tasks are aperiodic tasks where consecutive jobs are separated by a minimum inter-arrival time. Figure 11.1 depicts an example of scheduling three Liu and Layland tasks on a single processor.

11.1.1.2 Multiframe Model

In 1996 Mok and Chen [278] introduced the multiframe task model as a generalization to the periodic task model of Liu and Layland. A multiframe task τ

is described as a pair (T, C) much like the basic sporadic model with implicit deadlines, except that $\mathbf{C} = (C_0; ...; C_{N-1})$ is a vector of different execution times, describing the WCETs of N potentially different frames. The task generates an infinite succession of frames; the ready times of consecutive frames are at least T time units apart, the execution requirement of the i'th frame $(i \leq 0)$ is $C_{i \, mod \, N}$, i.e., the worst-case execution times cycle through the list specified by vector $\mathbf{C}$.

11.1.1.3   Generalized Multiframe Tasks

Baruah generalized the previous model in [61] by introducing the generalized multiframe (GMF) task model. In this model the deadlines of frames are allowed to differ from the minimum separation; further, all the frames need not have the same deadlines, and all the minimum separations need not be identical. A GMF task $\tau = (T; C; D)$ with n frames consists of three N-ary vectors $[C_0, C_1, ..., C_{N-1}]$ for WCETs, $[D_0, D_1, ..., D_{N-1}]$ for relative deadlines and $[T_0, T_1, ..., T_{N-1}]$ for minimum inter-release separations. In this case, for some offset a, the parameters for the job are release time $r_{i+1} > r_i + T_{(a+i)mod N}$, execution time $c_i \leq C_{(a+i)mod N}$ and deadline $d_i = r_i + D_{(a+i)mod N}$.

11.1.1.4   Elastic Model

Another generalization of the Liu and Layland model has been proposed by Buttazzo in [83]. The idea behind this model is to consider each task flexible with given rigidity coefficient and length constraints. In particular, the computation is fixed while the period can be varied within a specified range. A task $\tau_i = (C_i, T_{i_0}, T_{i_m in}, T_{i_m ax}, e_i)$ is characterized by the WCET C_i, a nominal period T_{i_0}, a minimum period $T_{i_m in}$, a maximum period $T_{i_m ax}$, and an elastic coefficient $e_i \geq 0$, which specifies the flexibility of the task to vary its utilization for adapting the system to a new feasible rate configuration. The greater e_i, the more elastic the task. The actual period T_i can vary according to it needs from $T_{i_m in}$ to $T_{i_m ax}$.

11.1.1.5   Mixed-Criticality Model

In the previous models, we can assume that there is another parameter to consider: the priority of the job, that defines its importance. However, in some scenario, we are more interested on the importance of the task and on its criticality (i.e. whether it is a hard or soft task). In this case, there is a model whose concern is to take into account different levels of criticality. It is called mixed-criticality model and it was introduced by Vestal in [373]. A mixed-criticality (MC) job is characterized by a 4-tuple of parameters: $J_i = (r_i, d_i, i, c_i)$, where r_i is the release time, d_i is the deadline $(d_i \geq r_i)$, x_i denotes the criticality of the job (with a larger value denoting higher criticality) and c_i represents the WCET. An MC instance is specified as a finite collection of such MC jobs: $I = J_1, J_2, ..., J_n$.

11.1.1.6 Splitted Task

The splitted-task model [82] is thought for systems that allow limited preemption. In this model each task tau_i is split into m_i non-preemptive chunks (sub-jobs), obtained by inserting $m_i - 1$ preemption points in the code. Preemptions can only occur at the sub-jobs boundaries. All the jobs generated by one task have the same sub-job division. The kth sub-job has a WCET $q_{i,k}$ and the total execution time of the full job can be obtained as $C_i = \sum_{k=1}^{m_i} q_{i,k}$. In this model each task is characterized by a 5-tuple $\tau_i = (C_i, D_i, T_i, q_i^{max}, q_i^{last})$ where C_i is the WCET, D_i the deadline, T_i the period, q_i^{max} the length of the longest sub-job and q_i^{last} the length of the last sub-job. The two latter values are important for schedulability: q_i^{max} is the longest blocking time that higher priority tasks can suffer due to task τ_i ; q_i^{last} affect response time since when it starts it cannot be preempted and task τ_i executes till completion.

11.1.2 Graph-Based Models

11.1.2.1 DAG Model and RT-DAG Model

The Liu and Layland model and the three-parameter sporadic task model both assume that there is a single thread of execution within each task. Those models are quietly simple and do not take into account factors as the modeling of parallelism within individual tasks or precedence constraints. This was not a problem in the context of uniprocessor real-time systems since there were no possibilities of parallelism on a single processor. However, nowadays the trend is to implement real-time systems on multiprocessors multicore platforms; this fact has given rise to a need of models that are capable of exposing any possible parallelism that may exist within the workload, including precedence dependencies between different parts of the same task.

A general parallel model which deals with the fine-grained execution provided by current parallel programming paradigms is the sporadic directed acyclic graph (DAG) model, proposed in 1998 by Baruah [60]. In this model, tasks are represented by means of directed acyclic graphs, with a unique source vertex and a unique sink vertex. Each vertex of the DAG represents a sequential job, while the edges of the DAG represent precedence constraints between these jobs (see also Figure 11.2 as an example of three tasks representation using DAGs). In detail, a task is specified by a 3-tuple (G, D, T) where:

- G is the DAG specified as $G = (V, E)$, where V is a set of vertices and E a set of directed edges between these vertices. Each $v \in V$ represent a sequential job, which is characterized by a WCET $e(v)$ and a deadline $d(v)$. The edges represent dependencies between the jobs: if $(v_1, v_2) \in E$ then job v_1 must complete execution before job v_2 can begin its.

- a period T. A release of a DAG task at time-instant t means that all $|V|$ jobs $v \in V$ are released at time-instant t. The period denotes the minimum amount of time that must elapse between the release of successive tasks.

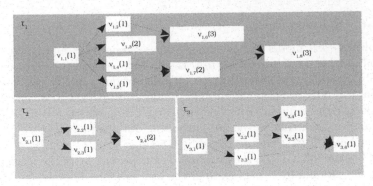

FIGURE 11.2: Representation of three DAG tasks.

- A deadline D. If a DAG task is released at time-instant t then all $|V|$ jobs that were released at t must complete by $t + D$.

Groups of jobs that do not have by precedence constraints may execute in parallel, whether there are processors available.

11.1.2.2   Digraph Model

There is a more general model w.r.t. DAG, introduced by Stigge in [349], that uses just directed graph and its called digraph real-time (DRT) task model. A DRT task T is described by a directed graph $G(T)$ with edge and vertex labeled as before. There are no further restrictions, any directed graph can be used to describe a task. However, every cycle in that model has to pass through the source vertex. Later, Stigge et al. propose in [350] an extension to this model that is called extended DRT (EDRT). In addition to a graph $G(T)$ as in the DRT model, a task T includes also a set $C(T) = \{(from_1; to_1; \gamma_1), ..., (from_N; to_N; \gamma_N)\}$ of global inter-release separation constraints. Each constraint expresses that between the visits of vertices $from_i$ and to_i , at least γ_i time units must pass.

11.1.3   Worst-Case Timing Analysis

Timing analysis frameworks determine an upper-bound on the WCET of a program or code fragment, to be given as input to a schedulability analysis modeled with one of the aforementioned models. Most of timing-analysis frameworks (such as [37]) analyze a program (part) that runs without interruption and in isolation, i.e., these tools do not consider all the interferences due to, e.g., access to shared resources, and to the co-existence in the system with concurrent tasks accessing the same computing cores.

Methodologies for WCET analysis can be roughly classified into four categories:

1. static techniques;
2. measurement-based analyses;

3. hybrid methodologies;

4. probabilistic methodologies.

While the first three methodologies are well affirmed, and recognized as equally important and efficient, the fourth one is more recent and thus fewer results are available as of today

Static analysis methodologies typically undergo three phases. At first, i) a **flow-analysis phase**, where a control flow graph is built for the application (typically, inside the code compiler or some specialized tool), and then we attempt to identify the so-called critical execution path, that is, the path(s) that leads to the higher execution time, in the worst case. Then, ii) the **low-level analysis** phase, where an estimate of the execution time is given for every atomic portion of the application code, such as assembly instruction, basic blocks, or small functions. Finally, a iii) **final WCET estimation** phase is carried on, where the derived flow (Phase 1) and timing information (Phase 2) are combined into a WCET estimate of the whole critical path, or parts of it. While these approaches are provably correct from the analytic viewpoint, they rely on an accurate estimate of the timing behavior of the underlying hardware platform, and, additionally, might incur in over-pessimistic WCET bound estimates when low-level effects such as penalties for concurrent memory accesses are put in the framework.

The traditional and most common method in industry to determine program timing is by measurements, assuming that "the real hardware is the best hardware model". The program is executed many times on the actual hardware, with different inputs and in isolation, and the execution time is collected. For those approaches, the main challenge is essentially to identify the set of input arguments of the application that leads to its WCET, and their drawback is that the real hardware platform must be available, which might not be always possible, and also means that this analysis is *not portable* across different systems.

A possible mix of static and measurement-based techniques provides safe WCET estimates, tighter than static approaches and sufficiently hardware-independent, hence portable. However, such as in measurement-based approaches, they are not provably accurate/exact, that is, it is uncertain whether the actual WCET has been captured or observed. The most representative tool for carrying such an analysis is RapiTime by Rapita systems [324].

11.2 Architecture

11.2.1 Hardware Model

After introducing a few basic concepts on real-time systems and how they are treated in traditional platforms that assume single-cores, we now discuss the main issues in deriving real-time guarantees and ensuring predictability within embedded heterogeneous systems where a host is coupled to a (possibly) many-core accelerators. We now introduce, as an example, a typical platform that we consider in this chapter, taken by a real use-case from the automotive domain, which can be found more specifically in modern advanced driving assistance systems (ADAS) and autonomous driving prototypes. At high-level of abstraction, the targeted hardware model features the following components: 1) CPU, 2) Accelerator, 3) Memory hierarchy (see Figure 11.3).

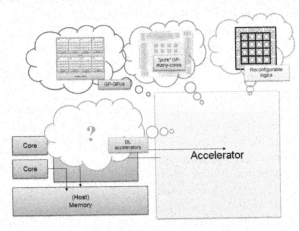

FIGURE 11.3: Target architecture.

As depicted in Figure 11.4, the Jetson TX2 is composed of two different CPU islands, a quad-core 1.9GHz ARMv8 A57 and a dual-core 2GHz ARMv8 Denver. Each of the cores in the A57 island integrates a private 32KB L1 data cache and a 48KB L1 instruction cache, while the Denver island cores integrate a 64KB L1 data cache and 128KB L1 instruction cache. Moreover, each of the islands has a 2MB L2 cache. The platform integrates an iGPU (integrated GPU[1]) Pascal-based architecture "gp10b" with 256 CUDA cores grouped within two streaming multiprocessor (SM), each of the SM in the SoC have a 64KB L1 cache sharing a 512KB L2 cache. Moreover, the iGPU and the core side share a 8GB LPDDR4 128bits DRAM able to reach an ideal bandwidth close to 60 GB/s [90]. The GPU integrates two major components that are the responsible of the different GPU functions, those are: 1) Execution engine (EE) that is in charge of performing the execution and 2) copy engine (CE) that is responsible of high-bandwidth memory transfers. The EE and CE can access central memory with a maximum bandwidth (close to 30 GB/s), which can saturate the whole DRAM (actual) bandwidth.

---

[1]since the TX2 has only one iGPU from now on we will call it GPU

11.2.2 The Issue is the Memory

The proposed architecture represents a powerful advance in the field of embedded computing systems, especially from the viewpoint of performance and programmability. Since the memory blocks are shared between the host and the accelerator subsystem, it is possible to completely remove the (costly) data copies to and from the accelerator itself, that are required in a "traditional" GP-GPU system, such as a desktop GPU[2]. The former approach is also known as *integrated GP-GPU*, as opposite to the latter, *discrete* GP-GPU. However, such an approach also has the main advantage of introducing serious interferences among the two subsystems.

These interferences greatly affect the timing analysis, which has to take in account a penalty for every memory access which is directly proportional to the number of computing cores competing for the bank, that is, potentially, all the cores in the system. If accesses to memory are not mediated, in the worst case, it is possible that every time a core needs to fetch data from the RAM banks (e.g., for a cache miss) it faces $N_{CORES} - 1$ times the latency for a single memory access. When applied to traditional timing analysis, this simple assumption causes extremely pessimistic, over-conservative worst-case timing bounds, that ultimately make the results inapplicable within a real application.

In [95] Cavicchioli et al. carried on a thorough analysis on a set of target architectures that resemble the one depicted here. Results are described hereafter, and make it clear how the interference due to memory contention can easily become critical, when a high number of cores comes into play, such as a many-core GPU accelerator.

Experiments. The target architecture considered in [95] is the ancestor of the Tegra TX2 board, namely the Tegra X1 [287], whose main differences being the smaller host core complex (four ARM cores vs. the 2 Denver + four A57 cores of the TX2), and the less powerful GPU, from the Maxwell second-generation "GM20b" with 256 CUDA cores grouped in two streaming multi-processors (SMs) (wrt the Pascal generation of the TX2). This results in a lower declared performance, i.e., 1 TFlops single precision peak performance drawing from 6 to 15 W). Figure 11.4 shows the most relevant components and notable contention points. The first contention point is represented by the LLC for the CPU complex. This cache is a 16-way associative cache with a line size of 64B. Contention may happen when more than one core fills the L2 cache evicting pre-existing cache lines used by other cores. Another contention point for LLC is indicated as point 3 in Figure 11.4, and it is due to coherency mechanisms between CPU and iGPU when they share the same address space. In the GK20a iGPU, such coherence is taken care in an unclear and undisclosed software mechanisms by the (NVIDIA proprietary)

---

[2]This is often referred to as "zero-copy" scheme, or, in NVIDIA terminology, *Unified Virtual Memory (UVM)*, because both CPUs and GPUs share the common (virtualised) address space.

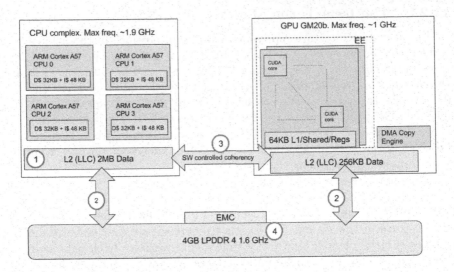

FIGURE 11.4: Hardware architecture.

GPU driver. Hardware cache coherence mechanisms take place only at CPU complex level. The remaining contention points (2 and 4 in Figure) are represented by memory bus and EMC for accessing the underlying DRAM banks. Such contention is of uttermost importance as it is caused by parallel memory accesses of single CPU cores and the iGPU. Interested reader can refer to [179] for finer grained discussions regarding the effect of bank parallelism in DRAM devices.

In order to stress the memory system under different workloads, the following (synthetic) tests have been performed:

Test Case A: intra CPU complex interference.

- *A1:* the observed core reads sequentially within a variable sized working set (henceforth *sequential read*), while the other cores are interfering sequentially (henceforth, *sequential interference*). Each iteration performs a *memcpy* of 100MB, so to involve every element within the CPU complex memory hierarchy (see also Figure 11.5).

- *A2:* the observed core reads with a random stride within a variable-sized working set (henceforth *random reads*), while the interfering cores perform sequential interference (see also Figure 11.6).

- *A3:* the observed core reads sequentially, while the other cores iteratively read 64B with a random stride within a 128MB array (*random interference*). The array size has been chosen to statistically prevent fetching already-cached data (see Figure 11.8).

- *A4:* the observed core performs random reads, while the interfering cores performs random interference (see Figure 11.9).

  Test Case B: iGPU interference to CPU.

- *B1:* the observed core reads sequentially, while the GPU accesses memory according to different paradigms: i) launching a copy kernel[3] between GPU buffers; ii) launching a copy kernel involving unified memory located buffers (CUDA UVM for X1 and K1, OpenCL SVM for the i7); iii) zeroing a device buffer(see Figure 11.7).

- *B2:* the observed core reads randomly, while the GPU activity is the same as B1 (see also Figure 11.10).

  Test Case C (CPU interference to iGPU).

- *C1:* the GPU accesses memory according to the different paradigms detailed in test case B, while the host cores perform the interfering patterns described in A1 (see also Figure 11.11 for the relative execution time comparison).

- *C2:* same as above, but host cores perform the interfering patterns described in A3 (see Figure 11.12 for the relative execution time comparison).

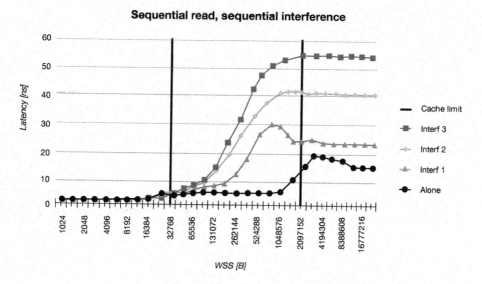

FIGURE 11.5: Test: A1.

---

[3]A copy kernel is a GPU program that performs an element-wise data copy between two buffers.

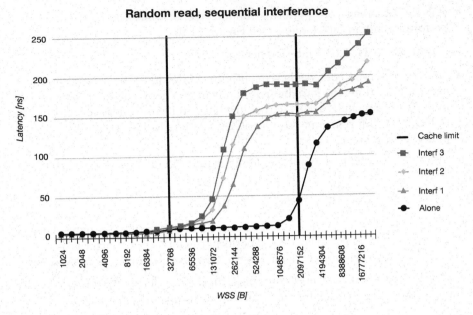

FIGURE 11.6: Test: A2.

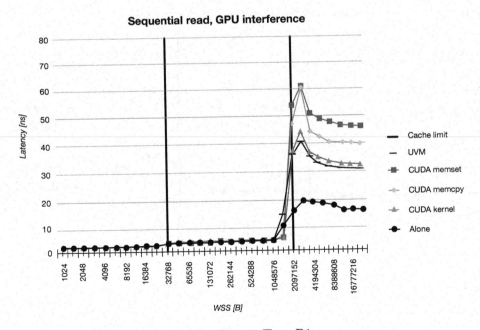

FIGURE 11.7: Test: B1.

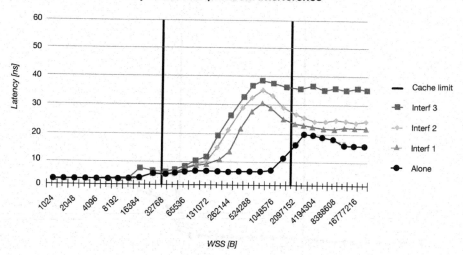

FIGURE 11.8: Test: A3.

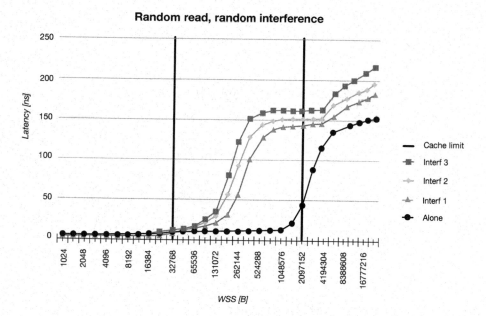

FIGURE 11.9: Test: A4.

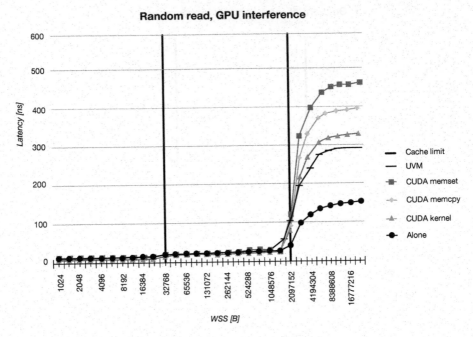

FIGURE 11.10: Test: B2.

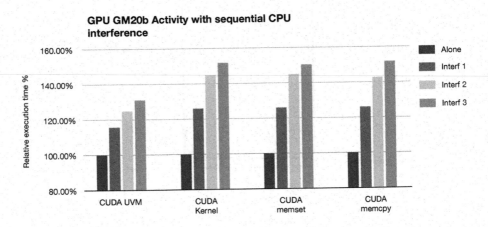

FIGURE 11.11: Test: C1.

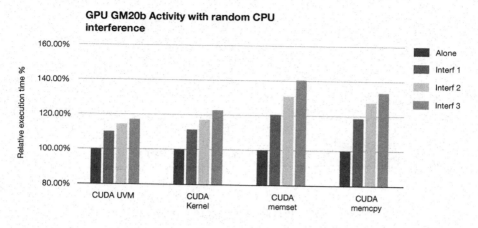

FIGURE 11.12: Test: C2.

11.2.3 Approaches to Mitigate this Effect

Modern heterogeneous architectures [289, 383, 224] typically feature accelerators composed of hundreds-to-thousands of computing elements. As explained in the last section, this rich degree of parallelism introduces severe performance bottlenecks due to access to shared resources (we explained this using shared SRAM banks as an example). Recently, several approaches have been proposed to mitigate this aspect, especially targeting commercial-off-the-shelf systems (COTS), that is, with no modifications on the existing hardware, but rather relying on pure software solutions, which are typically implemented at the OS, or hypervisor level (see also Section 11.3).

As an example of how this can be done in a multi-core system, we first introduce a very promising approach called the single-core equivalence, which mitigates contention effects due to shared memory and busses. Then, the following sections show how the accelerator itself (in our case, a GP-GPU) can be effectively abstracted and shared among multiple real-time tasks so that the overall predictability of the system is preserved.

The single-core equivalence (SCE) framework was introduced by Pellizzoni et al. [266]. It mitigates effects due to shared memory contention in a multi-core system by abstracting the underlying hardware and providing spatial and timing isolation to running tasks. Figure 11.13 gives a pictorial overview of the framework, highlighting the main sources of contention in a multi-core system made of four cores, with a shared L2 cache, a shared bus interconnect, and shared RAM. Starting from the top to the bottom, it shows how the contention effect on the shared caches can be mitigated using cache coloring/lockdown techniques, where one or multiple cache lines are exclusively assigned to

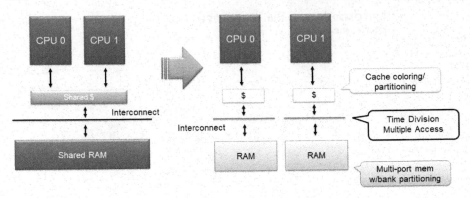

FIGURE 11.13: Single-core equivalence.

software tasks. This removes unwanted evictions, increasing the spatial isolation of the system, hence, its predictability. Contention on the shared interconnection can be either mitigate i) using high-bandwidth cross-bar with dedicated core-to-bank wiring, or ii) with more "traditional" techniques based on time sharing. The latter is also referred to as time-division multiple access (TDMA), and it can be included directly in the scheduling framework, as shown in [81]. Finally, there are several ways to reduce contention on shared RAM banks. One possible solution is to feature a special hardware configuration with per-core dedicated banks and interconnectivity, as it happens in the Kalray MPPA chip series [224]. Of course, such a powerful mechanism must be exposed to the higher (OS or user code) levels of the application stack, where memory is often managed by-hand. Pellizzoni [266], for instance, proposed *PALLOC*, an OS-level mechanism/primitive to share memory bank exclusively among a subset of running tasks, or to a single task. Finally, the probably more interesting approach is to share the available memory bandwidth among different computing cores and task, in a time-sharing fashion. This technique is often referred to as *memory throttling*, and can be successfully applied to any shared resource in the system (e.g., peripherals), and it's hardware-agnostic. We describe it in the next section, introducing the MEMGUARD mechanism developed in the Hercules project.

11.2.4 MEMGUARD

MEMGUARD is a memory bandwidth regulation mechanism for the multicore platforms, developed in the Hercules project. It allows sharing the bandwidth among the computing cores in the way that any single core is guaranteed to access the memory bus for a predefined time over a specified period. The software bandwidth controller monitors the memory access times for each core and if any core has exhausted its budget within its current period, the memory accesses from this core are throttled. The regulation period is the same for all the cores and the budgets are replenished synchronously at the beginning of the period. The tasks are statically partitioned among the processors.

Memory is a global shared resource and can be accessed by all cores incurring the same latencies of memory accesses for each core. Processors synchronously wait for every prefetch instruction caused either by cache miss or by explicit prefetch instruction (in reality it may concurrently execute out-of-order instructions). Given L as the maximum delay on a DRAM transaction for the core under analysis, each core reserves memory bandwidth represented by memory access budget Q_i for every period P. Regulation period P is a system-wide parameter which is the same for all the cores. The budgets are replenished synchronously at the beginning of the period. The sum of all cores budgets is less than the regulation period time:

$$\sum_{i=1}^{n} Q_i \leq P. \tag{11.1}$$

With such a mechanism, it is possible to compute the WCR iteratively starting from the WCR of single tasks. For reasons of space, we only report the final steps of the mathematical framework, first proposed by Yao, which is introduced and discussed in [388, 92, 363]. The WCRT of the task set is computed considering the continuous execution as one single task instance with the following parameters:

$$C^m = \sum_{j \in hep(i)} \left\lceil \frac{R_i}{T_j} \right\rceil C_j^m \tag{11.2}$$

$$C^e = \sum_{j \in hep(i)} \left\lceil \frac{R_i}{T_j} \right\rceil C_j^e \tag{11.3}$$

The response time of task τ_i is given by:

$$R_i = C_i + \sum_{j \in hep(i)} \left\lceil \frac{R_i}{T_j} \right\rceil C_j + stall(C^m, C^e, Q, P, m). \tag{11.4}$$

The formula can be solved iteratively.

11.2.5 SIGAMMA

In integrated system-on-chips (SoC), safety-critical tasks with tight deadlines are traditionally executed at the host side, offloading parallel kernels with a lower criticality to the GPU. Thus, as already stated, memory contention represents a significant threat for predictability and timing analysis. The problem is magnified if the considered SoC features a high performance GPU able to saturate the available memory bandwidth, increasing WCET up to 5 times due to the concurrent execution of a memory-intensive GPU application in modern integrated devices such as the ones previously described.

Such a significant latency increase motivates the need for a proper arbitration mechanism of memory accesses between host and device. In [92], Capodieci et al. presented *SiGAMMA* (transl. *SiΓ*), a server-based mechanism that acts as a memory arbiter between CPU and GPU, moderating the penalties due to the concurrent memory accesses by GPU engines. CPU tasks are assumed to be compliant with a predictable execution model (PREM) that separates memory phases from purely computation phases, as explained in [309]. Such an execution model allows predictably bounding the delays due to the concurrent access to shared memory resources by real-time tasks in multi-core environments.

While being inspired by the above-mentioned memory arbitration mechanism, *SiΓ* presents a set of distinguishing features:

- It uses a dynamic, event-based approach that allows for a better utilization of the memory bandwidth also for dynamic task sets.

- It does not rely on hardware counters, that, depending on the SoC implementation, may be too coarse grained to be used in real-time settings, especially for general-purpose embedded devices.

- It allows throttling GPU-side activities to limit their memory interference in heterogeneous SoCs using a novel and flexible mechanism, that is applicable also to closed-source architectures in which the possibility of modifying drivers is extremely limited.

A PREM task is composed of three phases: load, compute and unload. During the first phase, the task loads all data required for the subsequent computation phase in LLC; the next phase computes the pre-fetched data from the local memory, without needing to access system DRAM; in the last phase, elaborated data is flushed into main memory and/or copied in output buffers. Critical tasks having working set sizes larger than the LLC may be conveniently split into multiple PREM-ized iterations.

The *SiΓ* approach adopts a memory server that arbitrates the access to shared DRAM. Any host thread wanting to start a memory or computation phase needs to first access this server to notify the duration and nature of the following phase, namely, a memory phase (CPU_MEM), a computation phase (CPU_COM), or no operation (NO_OPER). The first one is used to signal to the server that a critical CPU thread is asking to exclusively access system DRAM; the second one informs the server that the CPU is about to undergo a compute phase and it will not resort to DRAM accesses; the latter one informs the server that CPU threads PREM iterations are over, hence releasing any locking privileges for system memory.

The GPU thread has read-only access to the server. Before performing any operation, it contacts the server to monitor the ongoing phase at the CPU side. If the CPU_MEM signal is active, the server does not allow any GPU operation. In this case, the server returns the remaining time of the memory phase, after which the GPU thread will contact the server again. If the host is instead in CPU_COM state, the server computes the remaining time of the computation phase, and allows the GPU thread to execute for a corresponding amount of time.

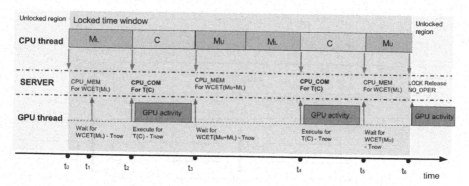

FIGURE 11.14: Timeline of an example sequence of interactions between a real-time CPU thread and a best-effort GPU task.

As shown in Figure 11.14 *SiΓ* is able to efficiently orchestrate CPU and GPU memory accesses. By means of a mediation effect enabled by a memory arbitration server, the memory contention due to GPU engines is drastically reduced, allowing host-side real-time tasks to access memory with little to no performance loss, even when executing in parallel with GPU activities. Find the complete results and discussion on the GPU efficiency loss in the original paper.

11.3 Software Support and Programming Models for Future RT Systems

Till now, we discussed cutting-edge research methodologies to mitigate the contention on shared resources (in our case, we discussed the memory banks) in multi- and many-core embedded systems. However, in order to build a fully operational real-time platform, the whole software stack must be modified to support those mechanisms in an efficient manner. We now discuss how this is done in the Hercules project. Finally, we briefly discuss how existing programming languages, tools and frameworks for such systems can be extended to support architecturally heterogeneous real-time systems. We target a multi-OS software stack (depicted in Figure 11.15) that supports different levels of criticality in the same system, in a platform such as the Tegra X2 where, as explained, an asymmetric host complex made by two Denver cores and four ARM A57 cores is coupled to a GPU. This platform supports on the same architecture both real-time AUTOSAR-like applications [306] running on top of ERIKA Enterprise [148, 164], and non-real-time (but high-performance)

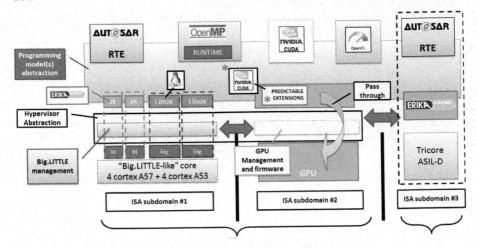

FIGURE 11.15: Software stack architecture.

computations performed partly in the Big.LITTLE-like subsystem and partly in the many-core accelerator. In addition to this requirement, it aims to support ISO 26262 certifications [156] of some of the safety-critical parts. Then, it becomes mandatory to guarantee proper isolation between the hard real-time parts and the rest of the system, in order to obtain the freedom from interference required by the standard in part 6, annex D.

For these reasons, the hard real-time subsystems have been properly "isolated":

1. For high levels of safety requirements, the project integrates an AUTOSAR subsystem supporting an external ASIL-D compliant CPU, such as the Tricore AURIX [207].

2. For lower levels of safety requirements, the real-time subsystem is integrated in the Big.LITTLE cores. **hypervisor** is used to separate and isolate the real-time subsystem (run under the ERIKA Enterprise kernel) from the rest of the system.

Both spatial and timing isolation is achieved using an infrastructure, based on the minimal open-source Jailhouse hypervisor for the host side. **Jailhouse** [344], developed by Siemens, is a Linux-based hypervisor oriented to real-time. Jailhouse isolates the virtual machines in small cells with few lines of code (13513 written in C), removing all of the unnecessary features (e.g., hooks for diagnostic tools), and schedules the virtual machines by pinning them to the computing cores. It also allows running bare-metal applications alongside to Linux.

For the accelerator part, unfortunately, "hiding" one or multiple GPUs under a hypervisor introduces a serious performance penalty for crossing its software layers, which ultimately might compromise the advantage of many-core acceleration. For this reason, a common solution is to provide a so-called *pass-through* mechanism for the CUDA drivers, which are allowed to bypass the virtualization layers and direct access to the device. This mechanism is represented by the arrow in Figure 11.15 that directly accesses the GPU, and it's currently supported on a limited set of GPUs, and for specific drivers[4]. A number of hypervisors and virtualization schemes exist for GPUs. Interested reader might refer to [353] as a good survey.

For the host part, the choice of ERIKA Enterprise makes it possible to run traditional AUTOSAR applications on top of an open-source implementation. ERIKA Enterprise [148] is currently the only open-source OSEK/VDX certified operating system, and it implements some of the extensions specified in the AUTOSAR OS standard. This opens the possibility to run traditional automotive applications with minimal or no change. ERIKA Enterprise runs on the Big.LITTLE subsystem, and it already supports directly the Tricore AURIX architecture.

11.3.1 RTOS Extensions: RT-Linux

Linux is the best candidate to support the many-core programming models described in next Section, opening the potential of seamless integration between the Big.LITTLE subsystem and the many-core infrastructure used as an accelerator. While, originally, Linux was developed as a best-effort OS, there is recent great interest in supporting real-time property, that gave birth to several extensions to the scheduler and driver modules that are now mainline for the last few years. We now cite the two most relevant.

SCHED_DEADLINE [248, 380] was originally born within the context of the EU FP7 ACTORS project [69]. It provides a scheduling policy suitable for real-time applications by implementing a resource reservation algorithm in the Linux kernel. This algorithm, named Constant Bandwidth Server (CBS [Abe98]), implements resource reservations over an Earliest Deadline First (EDF) scheduler. Resource reservations [Mer93] is an effective technique for providing temporal isolation, which allows using real-time scheduling even in GPUs. The basic idea is to assign each real-time task a "reservation" consisting of a "runtime" (called "budget" in the real-time literature) and a "period". The scheduler will guarantee that the task will execute at most for a time equal to the "runtime" every "period". If the task tries to execute for a longer time, then the scheduler throttles it (avoiding selecting it from execution) until the end of the current reservation period. In this way, each task is constrained to not use more than its reserved CPU share (called "bandwidth" in the real-time literature). Although referred with the name of "CBS", the algorithm actually implemented in the original SCHED_DEADLINE is HCBS

---

[4]For instance, NVIDIA published [286] a list of applications which are certified for this technology (called NVIDIA Grid).

(i.e. CBS with Hard Reservation [253]). In fact, if a real-time task tries to execute for more than its share, it is blocked, even when no other tasks are ready for execution. In the scheduling terms, this means that the algorithm is not "work conserving", meaning that it does not allow fully exploiting the system resources; it would be therefore desirable to allow the real-time tasks to get a bigger CPU share whenever the system is idle or underutilized.

The Greedy Reclamation of Unused Bandwidth (**GRUB** [253]) algorithm aims at introducing the reclaiming property to the CBS-class of algorithms, preserving the same semantics and kind of real-time guarantees. It introduces a dedicate state entered by a task that blocks and whose share cannot (yet) be reclaimed. An in-depth description of the algorithm along with a few examples can be found in the real-time literature [253] or as a patch to the Linux Documentation.

11.3.2   Extenting High-Performance Programming Models: OpenMP

OpenMP [120, 295] is increasingly being supported by the newest high-end embedded many-core processors. Despite the lack of any notion of real-time execution, the latest specification of OpenMP (v4.0) introduces a tasking model that resembles the way real-time embedded applications are modeled and designed, i.e., as a set of periodic task graphs. This makes OpenMP4 a convenient candidate to be adopted in future real-time systems. OpenMP defines *task-scheduling points* (TSP) as points in the program where the encountering task can be suspended and the hosting thread can be rescheduled to a different task. As a result, TSPs divide task regions into task parts (or simply parts) executed uninterrupted from start to end. The code listing shown in Figure 11.16 identifies the parts in which each task region is divided, e.g. T_0 is composed of $part_{00}$, $part_{01}$, $part_{02}$ and $part_{03}$.

The execution of an OpenMP program has certain similarities with the execution of a DAG-task as defined in Section 11.1.2.1: (1) the execution of a *task part* in the OpenMP program resembles the execution of a job in V for which WCET estimation can be derived; (2) the edges E in the DAG model can be used to model the OpenMP *depend* clause, which forces tasks not to be scheduled until all precedence constraints are fulfilled. The rightmost part of Figure 11.16 shows the OpenMP-DAG obtained by the example program presented in the listing on the left. Tasks parts are the nodes in V and the TSPs encountered at the end of a task part (task creation or completion, task synchronization) are the edges in E, The figure distinguishes three different types of edges: control flow dependencies (dotted arrows) that force parts to be scheduled in the same order as they are executed within the task, TSP task creation dependencies (dashed arrows) that force tasks to start/resume execution after the corresponding TSP, and TSP synchronization dependencies (solid arrows) that force the sequential execution of tasks as defined by the if clause, the dependent clause and the task-waits synchronization con-

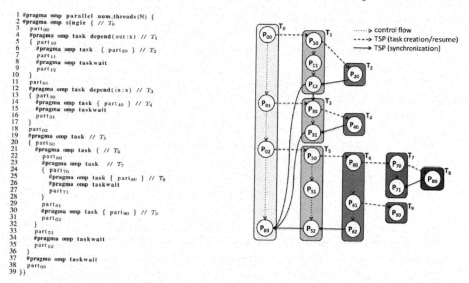

```
1  #pragma omp parallel num_threads(N) {
2  #pragma omp single { // T0
3     part00
4  #pragma omp task depend(out:x) // T1
5  { part10
6  #pragma omp task  { part20 } // T2
7     part11
8  #pragma omp taskwait
9     part12
10 }
11    part01
12 #pragma omp task depend(in:x) // T3
13 { part30
14 #pragma omp task { part40 } // T4
15 #pragma omp taskwait
16    part31
17 }
18    part02
19 #pragma omp task // T5
20 { part50
21 #pragma omp task { // T6
22    part60
23 #pragma omp task  // T7
24 { part70
25 #pragma omp task { part80 } // T8
26 #pragma omp taskwait
27    part71
28 }
29    part61
30 #pragma omp task { part90 } // T9
31    part62
32 }
33    part51
34 #pragma omp taskwait
35    part52
36 }
37 #pragma omp taskwait
38    part03
39 }}
```

FIGURE 11.16: Example of OpenMP application based on tasks (left), and its DAG representation (right).

struct. All edges express a precedence constraint. This example is taken by a work [340] developed during the FP7 P-SOCRATES [364, 314], project, which greatly inspired the Hercules project by proposing predictable extensions to HPC programming models.

11.3.3   Extending Accelerator Programming Models: CUDA

To develop real-time applications that will benefit from GPU acceleration, a real-time compliant scheduler of the accelerator itself was needed. The first implementation of preemptive EDF on GPU with Constant Bandwidth Server (CBS) has been implemented by Capodieci et al. in [91]. The implementation is based on a software RunList Manager (RLM) that acts as a privileged virtual machine in the NVIDIA Hypervisor, managing and scheduling the tasks to be submitted and the preemption signals to be sent.

The authors here also presented API extensions for OpenGL and CUDA to describe Real-time task properties and to pilot the RLM scheduler. The minimal CPU-to-GPU submission mechanism for these novel APIs involves minimal driver interactions and validation procedures, therefore minimizing the impact of CPU-side delays during command submission. This is in contrast with the traditional APIs (e.g., CUDA and OpenGL) and respective programming models, in which commands are constantly streamed from the CPU to the GPU, with each API call being validated at driver level. Such a paradigm not only constitutes an additional threat to predictability, but it also makes it impossible for CUDA applications to define a concept of batched command submission or their DAG representation.

The extensions are basically additional user space runtime OpenGL and CUDA functions that internally trigger appropriate messages and signals to the RLM. Graphic applications are intrinsically batched; on an application-level perspective, this resulted in the creation of an OpenGL API call able to inform the RLM about the rendering WCET and the desired target frame-rate.

For CUDA compute applications, instead, commands are not batched, as there is no equivalent concept of frame boundary. In order to create batches of commands, the authors introduce two additional CUDA runtime API calls: *cudaStreamDeadlineBegin* and *cudaStreamDeadlineEnd*.

These API calls allow binding different batches of commands within different CUDA streams, where a CUDA stream is a software abstraction of a queue of commands which are executed in the order they are inserted into the stream. Therefore, the API extension allows to identify task boundaries where to define scheduling parameters (period, budget and relative deadline) that will be then associated by the RLM to all the commands included between the code block of *cudaStreamDeadlineBegin* and *End*.

Scheduling parameters are inserted as input arguments for *cudaStreamDeadlineBegin*. More specifically, the input arguments for the added function calls are:

- *cudaStream_t s* : the CUDA stream in which enqueue the commands to schedule.
- *uint32_t D_r* : the relative deadline of the batch of commands [μs].
- *uint32_t B* : the budget of the batch of commands [μs]
- *int32_t P* : the period of the batch of commands [μs].

Work completion notification from CPU side to the GPU is implemented through *cudaStreamDeadlineEnd*, which is a wrapper to the CUDA standard runtime function *cudaStreamAddCallback*. This function registers an asynchronous callback to notify the RLM when the previously enqueued operations of the CUDA stream are completed.

11.3.4 Industrial Modeling Frameworks: The AMALTHEA Case

In the automotive domain, AUTOSAR (Automotive Open System Architecture [3]) is an open-standard automotive software architecture for vehicle electronic control units (ECUs). To provide software scalability and portability, the AUTOSAR model is divided into different layers. AUTOSAR software components act as car functions and are implemented independently of the underlying hardware. In this way, it is possible to use the same software components in different ECUs. AMALTHEA is an open-source standard to model the full deployment of a project running AUTOSAR software components. It includes key properties of the computing system architecture, such as the number of cores and NUMA memory banks, and, thanks to its simple XML-based layout, it is extremely useful to automotive engineers in the V&V process, and it is being adopted by leading Tier 1 companies such as Bosch Gmbh.

Model-based design is a common paradigm to create abstract representations of industrial applications. Such a process is typically automated through the so-called Model-to-Model Transformation (MMT) and Model-to-Text (M2T) transformation. In a nutshell, the code generation process receives the model as input, and then, if necessary, it converts it into a different model with the MMT process. This can be, for instance, the conversion of a UML model into another model that better captures certain properties of the target system. Finally, the model is transformed into a lower level code representation (e.g., C/C++) through the M2T procedure. Figure 11.17 shows this process.

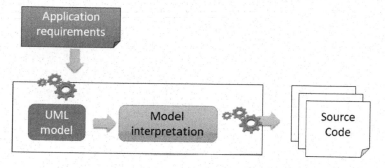

FIGURE 11.17: Code-generation process.

The architectural model by Amalthea is capable of modeling multi-core systems based on shared memory and a system interconnection such as cross-bars/predictable busses. However, it currently lacks support for expressing accelerator co-processors, such as GP-GPUs or FPGA logics. Extending those models represents a key challenge for including hardware heterogeneity within industrial frameworks.

12

Future Challenges in Heterogeneity

A. Scionti, O. Terzo, and F. Lubrano

LINKS Foundation – Leading Innovation & Knowledge for Society, Torino, Italy

C. Pezuela

Atos, Madrid, Spain

K. Djemame

University of Leeds, Leeds, UK

CONTENTS

The continuous demand for performance fed by emerging machine learning (ML) and Big-Data (BD) applications, as well as the growing technological limiting factors are pushing system architects to ever more exploit "heterogeneity" in their solutions. Besides performance, energy consumption and programmability represent hard constraints which limit the complexity (and thus possible functionalities) achievable by modern systems. To overcome such obstacles, in recent years, architectural specialization has been massively explored in order to ever increase performance per Watt, while it will represent the key for keeping the pace dictated by the Moore's law. Such architectural *diversity* will emerge at different levels and scales in designing new computing systems, posing new challenges to be addressed. Cloud architectures are ever more diffused and decentralized and include an ever large fraction of processing "at the edge" (i.e., Fog/Edge computing [262]). Orchestrating the resources, as well as the interaction of such sparse elements becomes more challenging, and further extends the concept of "heterogeneous computing architectures". Packing together an increasing number of transistors in a single die is becoming more difficult. To work around such difficulties, designers are looking other ways for building ever complex processors. For instance, chiplet-based systems allow to combine specialized and optimized architectures which are hosted on dedicated modules, eventually manufactured using different technologies and placed on a common substrate. Similarly, computing nodes being part of large-scale computers integrate many forms of acceleration, ranging from many-cores to reconfigurable fabrics. Unlike in the past, specific application domains are driving the creation of modern architectures (ranging from single chip designs to entire clusters), as strongly remarked by machine learning and Big-Data applications. Although very welcome, such vast variety of hardware elements will pose challenges in terms of management and programmability. Not only processing elements will become more heterogeneous in the future, but also memories, storage and interconnections will be too. In such context, operating systems, orchestration tools (largely used to manage large-scale infrastructures) and middleware will be forced to adapt their resource assignment policies and schemes by taking into account hardware diversity. Similarly, programming languages will necessary have to integrate more sophisticated mechanisms to efficiently exploit underlying hardware and free programmers from manually targeting different architectures, letting them focusing on the applications. For instance, abstraction and virtualization mechanisms will have to be included to enable a more flexible and dynamic utilization of the resources on an ever larger scale. As the hardware changes and becomes more differentiated, applications have to be, in general, rethought from their roots to fully exploit the innovative features and get benefit. To support programmers in taking advantage from new hardware, development frameworks must adapt to the changes. To this end, programming languages and development frameworks will evolve by including more and more capabilities, freeing the programmers from the need to extensively declare "how" to accomplish a specific goal.

The scope of this chapter is twofold. On one side, it aims at summarizing the material presented in the previous chapters. Specifically, through the description of the evolution of the hardware and software heterogeneity landscape, this chapter tries to highlight the link with specific research topics touched in previous chapters. On the other side, this chapter aims at marking both future challenges and opportunities that designers will face.

12.1   Introduction

Despite the continuous improvements made in manufacturing digital computers, there is still always a growing demand for more performance and capabilities. Such demand has different origins: from end user applications (here, intended as hardware and software) which require more capabilities and efficiency for running longer when powered by batteries, to Cloud providers which need to process a growing massive amount of data, to scientific applications which require the best-in-class performance. In order to satisfy this insatiable computing capability hunger, while being able to face all the challenges associated, "specialization" (heterogeneity) has been massively adopted. From chip architectures, to nodes and distributed systems, heterogeneity has become pervasive and multifaceted. Depending on the application context, specialization is used to achieve different goals. Also, achieving such goals requires a more general approach, where different technologies must be integrated.

This chapter provides an overview on recent evolution of this heterogeneous technology landscape. Although, design specialization is not new, its broader adoption requires more systematic (and holistic) approaches to be able to face new challenges, as well as to take advantages from new emerging opportunities [138]. The remainder of the chapter is organized as follows. Section 12.2 introduces an overview of the computing continuum, focusing on heterogeneity architectural aspects. Also, a focused (albeit not exhaustive) description of modern processing architectures and their evolution is provided (this includes processing functionalities and micro-architectural aspects, memory organization and interconnections)—see Section 12.3. Here, reliability aspects of modern systems will be also discussed. Section 12.4 focuses on programming aspects, while Section 12.5 moves the analysis at the higher level of abstraction, by considering the management of heterogeneous environments. Section 12.6 summarizes the content of the chapter.

12.2   The Heterogeneous Computing Continuum

Digital technologies are becoming ubiquitous in our lives, from smart IoT sensors to large-scale data centers we are submerged by digital data. Cloud computing providers have recognized the need of adequately collecting, processing and storing such data. To this end, the concept of Cloud computing, generally used as synonymous of platforms running on data centers, has been redefined to include more elements distributed till the edge of the network, where IoT devices generally are placed. Thus, modern Cloud computing architectures contain hardware and software elements the span over the traditional boundaries of data center infrastructures. This progressive movement of Cloud technologies to include elements ever closer to the "edge" of the network enables data to be generated and consumed at different levels. Thus, data flow in a computing continuum, which is also becoming ever more heterogeneous than in the past.

Figure 12.1 depicts a modern computing infrastructure, where IoT devices (left) dynamically interact each other and with remote (Cloud) services (right). However, the continuous demand for low latency processing capabilities pushes service providers and manufacturers to distribute processing

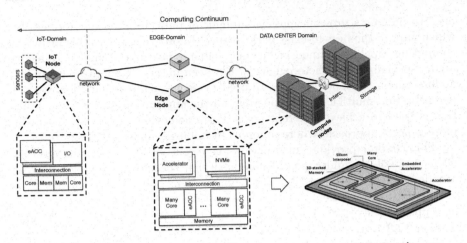

FIGURE 12.1: An overview of the heterogeneous computing continuum.

resources in-between the edge (IoT) and the large-scale data centers (DCs). Such resources are generally referred to as Fog computing (or simply Fog) resources. Similarly to hardware and software solutions found in DCs, also Fog resources explore heterogeneity, often integrating highly specialized hardware.

The growing pace at which data are generated poses new challenges in terms of capabilities and efficiency in processing and analyzing them at different levels of such architectures. In such context, one of the main revolution we are assisting in recent years is represented by artificial intelligence (AI). Among the various faces that compose this huge domain, machine learning (ML) and deep learning (DL) techniques have become popular. With the advancements in processing technologies (ever more powerful processors and dedicated accelerators), DL algorithms became largely used in many diverse applications, spanning from autonomous driving vehicles and enhancing photos on smart-phones to fast detection and classification of massive volumes of data. For instance, large popular Cloud providers (e.g., Facebook, Google, Microsoft, etc.) have to run daily such algorithms to ensure loading of appropriate contents in their respective social platforms. Chapter 9 described the implementation of a state-of-the-art Convolutional Neural Network (CNN—a subclass of the wider class of DL algorithms) on a heterogeneous low-power device. Chapter 8 remarked the strict relationship between heterogeneous systems and ML/DL algorithms, by studying the implementation of an image recognition application on Amazon Compute Cloud and comparing the obtained results with their counterparts running on modern CPUs, GPUs, and FPGAs. Running faster and efficiently such algorithms becomes important to support the growing rates at which data have to be ingested. The problem is exacerbated by the complexity of training these algorithms. While task inference (i.e., the classification of the input patterns according to internal weights previously learned through examples) can be efficiently performed on many different type of hardware, training phase is not. Training phase generally requires high-performance computing systems to be performed in a reasonable amount of time. This stringent requirement pushed the attention of both large Cloud providers, as well as of hardware manufactures to investigate more on dedicated acceleration platforms, thus further enlarging the spectrum of heterogeneity. Compared to CPU micro-architectures, such accelerators show larger (massive) parallelism (e.g., GPUs, Google TPUs) and some degree of adaptability (FPGAs), using simpler cores tailored for efficiently executing common operations used in DL algorithms. In other cases, the underlying organization of the accelerators is completely substituted with other structures (in some cases, the physical devices are based on a technology that is different from conventional CMOS one) that allow to mimic the functioning of biological neurons (i.e., neuromorphic architectures). ML and DL algorithms are at the basis of many analytic tools used for supporting the analysis of massive data. Although, modern processor and accelerator architectures offer high performance and efficiency, their capabilities need to be push further. Scalability and adaptability will become important aspects to ensure next generation computing system to run new and more complex algorithms. The former refers to the capability of specialized hardware to be integrated in much larger systems in such a way to outperform capabilities of the single system. While this already happens for CPU-based nodes, less work has been done for accelerators [161]. The latter is intended to provide capability of the hardware to run future algorithms without the need of costly silicon updates [153]. Similar requirements remain valid also for devices based on radically new approaches, if their adoption at scale is aimed.

Besides architectures for data processing, also data are becoming more and more heterogeneous, being generated by different sources and carrying different type of information. In this context, it is expected that ML/DL techniques will greatly improves the capabilities of extracting meaningful information from such data. Indeed ML/DL techniques are considered the basis of the fourth research paradigm. The first twos are represented by theoretical modeling of a phenomenon and the experiments to validate the theoretical model. Recently, the increased capabilities of (super)computers allowed for introducing the third paradigm: computer simulations can be used conveniently to predict the behavior of a studied phenomenon, as well as to easily verify the theoretical models. In this respect, DL/ML are used to extract useful information from raw data produced by experimentation or by simulations (fourth paradigm). Evolving ML/DL towards ever smarter techniques will allow computer systems to automatically propose theoretical models, plan simulations and experimentation to validate them, and extract the useful information from generated data. A step forward with respect to current state-of-the-art is required to enable ML/DL algorithms extracting automatically useful information. Programming frameworks and languages must evolve to let programmers expressing "what" it should be accomplished, rather than "how".

Also, traditional HPC community increasingly started to integrate ML/DL algorithms in complex simulation work-flows, as well as to integrate dedicated acceleration hardware in their system. Chapter 7 described a framework for easing the adoption and use of FPGA devices in the HPC context. Again, the influence of ML/DL in one domain contributed to its evolution towards much larger heterogeneity. In this context, future challenges will be represented by the capacity of gluing together hardware and software stacks that will be designed to achieve different goals (i.e., HPC domain requires high-performances using high-precision floating-point formats, while ML/DL algorithms are more bounded by memory requirements and use less accurate numerical representations). In such, the role of resource orchestrators (that is more common in Cloud environment) will be necessarily extended to cover a broader set of resources.

There are other two key elements being part of any computer (irrespective of its size and scale) that play a fundamental role in sustaining performance and keep the system working efficiently. On one side, data being processed must be stored. Memory is the basic block that implements such function. On the other side, data need to be moved across the computing system components, passing from one unit or node to another. This second functionality is provided by interconnections. Both memories and interconnections evolved over the last decades to embrace more heterogeneous approaches. Looking at the memory side, thanks to the improved manufacturing processes, ever larger memory devices can be integrated in modern chips. Static RAMs are organized in multiple hierarchical levels to better serve hungry processing elements. Main memory, traditionally based on DRAMs, increased its capacity, as well as encountering faster interfaces that have been made available to sustain higher data throughput. The biggest change happened since the introduction of massive storage devices (MSDs) (typically based on spinning magnetic disks) has been the large adoption, in almost every type of modern computer, of flash memories. Although, technological improvements allow for better performance and increased capacity at every new device generation, new challenges appear at the horizon. Scaling of flash-transistor size is reaching physical limits, and thus new approaches are required. Magnetic spinning disks, still used in large-scale systems (e.g., hyperscalers' data centers) thanks to their storage capacity, will require the adoption of new radical technologies to guarantee increasing storage capacity without compromising data retention reliability.

Networking also evolved in the last years to include more sophisticated interconnection technologies. Inside data centers, latency and bandwidth are still of major concerning parameters to look at (current Ethernet solutions range from 10 Gbps to 100 Gbps), although the implementation of additional "smart" functionalities received a strong impulse. With a growing demand for flexible mechanisms to manage the underlying infrastructures, also the networking stack takes advantage from principle of virtualization. Software-defined networking (SDN) allows treating networking elements as virtualized instances that can flexibly deploy across a variety of hardware, including standard servers. Similarly to the evolution process that led to the transition from virtual machines to containers running microservices, also in the networking domain single network functionalities became the main target for (hardware) acceleration [275, 222, 153, 358]. In this respects, network interface cards (NICs) became "smart", including dedicated ASICs (application specific integrated circuits), multicore processors and in some case reconfigurable fabrics. Protocols also changed to better support data transfers among heterogeneous nodes (e.g., between GPU and CPU nodes), as well as to support Cloud virtualization frameworks (e.g., OpenStack). In the context of HPC domain, dedicate interconnections are well established key elements of any HPC-infrastructure. However, also in such domain, evolution led to the adoption of different types of technologies to support different classes of traffic and workloads [70, 42]. Furthermore, thanks to the increased link speed, highly-dimensional folded topologies have been designed. As mentioned at the beginning of this section, modern Cloud infrastructures extend towards the edge of the network. An essential component to enable these large-scale systems working is the networking infrastructure. With the large amount of data generated by IoT/Edge devices, high-bandwidth low-latency links must be used. *5G* promises to make a big step forward, enabling ultra-wide bandwidth wireless communications, which are of interest for many mobile applications (e.g., autonomous driving cars, vehicle-to-infrastructure communications, video/game streaming, etc.). However, despite the many opportunities offered by 5G technologies, other challenges must be addressed. Whenever a full control of the network infrastructure is required, ad-hoc network must be used. Different technologies will be required also for enabling communication in rural/harsh environments [313], thus increasing the heterogeneity we will see in the whole infrastructures. Wi-Fi communications already offer large flexibility along with high transmission rates. However future challenges will come by the extensions of link distances without compromising the other features provided by the Wi-Fi standards.

12.3   Processing Architectures Evolution

Chapter 1 provides a comprehensive overview of modern processing architectures that are at the basis of general-purpose processors (GPPs) and processor accelerators (PAs). Besides the historical evolution of GPPs, which moved from earlier single-core in-order micro-architecture to modern out-of-order multi-cores, in recent years there has been large adoption of emergin PAs too. Being created for supporting visualization of complex 2D/3D scenes, GPUs evolved to more complex systems. Still being based on micro-architectural designs sporting thousands of simple efficient cores, GPUs embed ever more specialized ones. Indeed, starting from first experiments [80] in using GPU hardware for accelerating general purpose applications (e.g., matrix-to-matrix multiplication, etc.), progress has been made in integrating capabilities of supporting operations used in scientific domain (e.g., support for double-precision floating point arithmetic) [107, 216, 2]. With progress in manufacturing technologies, it was quickly possible to integrate large amounts of transistors, making GPUs more oriented to support scientific applications. Indeed, with the availability of massive amounts of parallel cores designed for performing mathematical operations with high throughput, GPUs became the preferred platform for several generations of supercomputers [366], as well as for many hyperscalers (e.g., Microsoft Azure [12], Amazon EC2 [1]). Recently, the explosion of ML/DL-based applications, along with the increasing demand for real-time realistic image visualization, introduced new challenges and changes in the GPU architectures. To better support ML/DL workloads dedicated cores [268] have been introduced: the hardware structure is optimized to perform matrix-multiply-and-accumulate operations (using small input matrices) in few clock cycles (see Figure 12.2). The same changes in design have been followed for embedded graphic cores [354]. Dedicated cores supporting ray-tracing techniques have been recently introduced, as the case of Nvidia core-RT architecture [114].

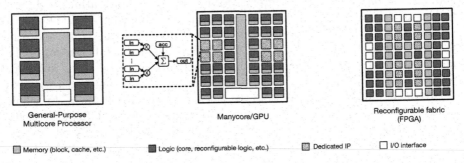

General-Purpose
Multicore Processor

Manycore/GPU

Reconfigurable fabric
(FPGA)

■ Memory (block, cache, etc.)   ■ Logic (core, reconfigurable logic, etc.)   ▨ Dedicated IP   □ I/O interface

FIGURE 12.2: Processor and accelerator heterogeneity.

Following the market trends, also GPP designers started to incorporate more specialization. Adding more complexity in the cores' micro-architecture extended support for low-precision arithmetic (as required by ML/DL applications), as well as vectorization. This is the case, for instance, of recent Intel processors [113] which introduced AVX-512 instruction-set extensions, mainly borrowed from previous generation of Intel Many Integrated Core (MIC) solutions [347]. Besides the large improvement in performance achieved by these new platforms, the trade-off between hardware exposed functionalities, programmability and support for applications residing in different domains (so, with diverse constraints), challenges in designing and manufacturing processors and accelerators quickly arise. Thus, in the race of building the fastest solution, specialization is often pushed to the limit, by creating ad-hoc designs.

Aiming at performing as the best in class systems for both training and inference phases of ML/DL algorithms, dedicated designs have been proposed, such as Tensor Processing Unit (TPU) [220] and Intel Neural Network Processor (NNP) [323]. Even more radical approaches have been studied, but their adoption as market products still remains far away [333, 163, 43, 122]. Among the others, architectures based on unconventional computing paradigms seem to be closer, since performance improvements are obtained by changing the way computations are managed, irrespective of the underlying physical implementation that still can be based on traditional CMOS [230, 336, 337].

The additional IP blocks embedded into the reconfigurable fabric to speed up specific operations, makes more difficult finding an optimal placement and routing of resources (which is fundamental to achieve performance and reduce energy consumption). Following the general market trend, and being a valuable platform for large-scale Cloud providers, FPGAs recently embedded a high number of such dedicated IPs, which cover a large fraction of the available transistors budget. For instance, many solutions from FPGA vendors contain multi-core blocks, ranging from simple 32-bit ARM cores to more powerful out-of-order 64-bit ones. Future systems are expected to further increase such design complexity by adding more diverse IP blocks [165], including those dedicated to complex mathematical operations largely used in ML/DL applications. Figure 12.2 shows such evolution of different processing platforms (general purpose processors and accelerators) towards more heterogeneous systems.

One of the key challenges in keeping the pace of such evolution is represented by manufacturing technology scaling, which is no longer achievable in an easy way. Since Dennard scaling has ended, manufacturers had to find other ways of scaling chip functionalities, e.g., by integrating more cores on the same die. Following this approach, however, is difficult since larger dies are more prone to defects. Despite large investments done to be in line with roadmaps and towards next manufacturing node sizes ($7\,nm$, $5\,nm$ and $3\,nm$), other ways are investigated to keep performance improvement running. To address such limitations, 2.5D and 3D stacking are considered the most viable solutions. The former allows placing separated dies (chiplets) on top of a common silicon interposer, which is used also to implement package-level interconnection. Chiplets can be manufactured with different technologies, according to their specific functionalities and avoiding scaling issues (e.g., interconnections subsystem scales worse than logic, thus recent designs placed interconnection box on a separated die [249, 284, 261]). The chiplet design approach also helps improve manufacturing yields and reduces investments since different modules can be reused. Further, 3D-stacking allows for exploiting vertical dimension to build complex design, also reducing the die-to-die communication latency, especially when monolithic approach is used.

Complex modular systems require efficient interconnections to avoid wiping away performance and energy efficiency gain. Over the years, different approaches to intra-die interconnections have been proposed, ranging from traditional buses to ring-base interconnections, to more efficient networks-on-

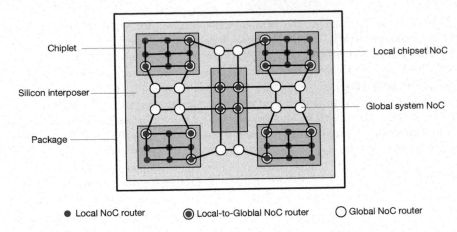

FIGURE 12.3: Heterogeneous interconnections within a modular chip: Chiplets are connected each other through a global NoC, internally they use local NoCs.

chip. These latter provide higher-throughput and better traffic management compared to other approaches; however, with the increasing number of sub-modules interconnected, their design became more complex too. Indeed, to better distribute network traffic, ever more complex topologies have been proposed. Although, traffic performance are still one of the key factors for successful designs, power consumption must be taken into account. In such respect, new challenges arose. Partial solutions came by adoption of hybrid topologies; on the other side, different ways of transmitting information have been considered. Among the others, silicon photonincs appears the most doable solution, although differences in the manufacturing technology with respect to standard CMOS slow down its adoption on a industrial scale. Future challenges will come by the extension of such interconnections to the package level, and further to the board level (see Figure 12.3).

Memory and memory hierarchy represent the third key element of any modern computer architecture. Memory performances (i.e., speed of read/write data) are fundamental for keeping the whole system performance at the desired level. Over the last decades, memory organization and also technology at its basis changed and became more heterogeneous. We can identify three clear separation levels in the whole memory hierarchy (from storage elements inside the chip to the massive storage on a node):

- **Chip level**. At this level, memory elements are generally based on SRAM cells to ensure fast data access. For instance, different cache levels of modern processors are basically large instantiation of SRAM cells. However, to cover the growing demand for fast accessing to large memory pools, DRAM-based solutions have been placed closer to the processor cores. To this end, 2.5D integration is used to place 3D-stacks of memory within the same package.

The logic layer of the memory stack (which integrates interface controller and decoding logic) is placed at the bottom of the stack; multiple layers of DRAMs are stacked on top of the logic layer offering higher bandwidth with respect to conventional DDRx architectures [307, 213] thanks to wider internal data paths. More bandwidth allows it to support concurrent requests in complex many-core designs. Slightly different architectures have successfully reached the market: hybrid memory cubes (HMBs) and high-bandwidth memories (HBMs) [233], multi-channel DRAMs (MCDRAMs) [347]. These type of memories are already used in high-end (GP-)GPUs and accelerators (e.g., Intel MIC, Intel Stratix10 [9], Xilinx Virtex Ultrascale [168]).

- **Node level.** Within a single node, large memory pools are provided by DRAM modules (DIMMs), which interface with the processor by means of dedicated data transfer protocols. Besides speed data transfer, one of the major challenges in this context is represented by the trade-off between memory storage capacity, speed in accessing data, and data retention. Technology scaling for DRAMs is reaching physical limits (it is becoming tough to scale down capacitors ensuring them to trap enough electrical charges to discriminate between 0 and 1), thus new technology are evaluated. NAND-based flash memories scale better than DRAMs and also keep their states when power supply goes off. However, their speed is far from that required by a good alternative DRAM candidate. Future non-volatile DIMMs will be based on different technologies to ensure enough read/write speed, which exploit thermally induced phase changes, formation of metallic conductive paths between electrodes, tunneling effects in a magnetic junction. Such new memory architectures will pose new design challenges, since interfaces between processor and memory modules must be revised in order to allow processing cores correctly address all the storage space. Interestingly, such larger memory space enables new computing paradigms to be implemented, such as the case of *in-memory computing* [173]. Indeed, loading huge amounts of data "in-memory" makes it more effective to move computations where data reside. To this end, simple operations can be performed directly on the memory cells, allowing to save large fraction of energy. Making such paradigms effective at scale, implies to face and solve challenges related to integration within the full software stack, and also enabling programming languages to express easily how such computations should be performed.

- **System level.** At system level, storage capability is more pressing rather than pure performance. Over the years, traditional magnetic spinning disks (HDDs) have been substituted by more efficient (with higher performance) solid-state drives (SSDs). Based on NAND-flash devices, SSDs offer superior performance, better energy efficiency compared to HDDs. However, when storage capacity per disk unit is analyzed, HDDs still provide more space. Hyperscalers, as well as large-scale infrastructures are always demanding storage space, and since data are managed in large chunks, HDDs are the preferred solution. Keeping the pace of storage demand, pushes HDD manufacturer to the limits. To address such challenges, new technologies are investigated, ranging from laser/microwawe-assisted heads to patterned media.

Despite the huge opportunities offered by such new design approaches, there are also new challenges to solve. Whenever two different dies must be interconnected, communication interface must be carefully designed in order to avoid signal degradation, and thus performance losses. However, a major issue in these new approaches is represented by difficulties in reducing the size of vertical links (through silicon via TSVs) while ensuring correct alignment between dies. Besides architectural challenges, one of the most difficult challenges to address in such scenarios (with the continuous growing complexity) resides in the way programmability and productivity are guaranteed. Squeezing performance from the underlying hardware is not the real problem; conversely, extracting performance still maintaining a clear and simple programming interface (i.e., instruction set architecture with a certain degree of generalization) is challenging. The more specialized the architecture is, the more effort will be required to the compiler tool-chain to map basic operations on the underlying hardware structures. The challenge is exacerbated by the growing number of developing frameworks that should be targeted (e.g., OpenCL [15], TensorFlow [34], Caffe [215], Theano [44], Torch [110], etc.). When moving to reconfigurable hardware (i.e., FPGAs) mapping tasks become even more complex, since high-level statements must be translated into a data flow graph, which in turn must be associated to underlying logic blocks. Furthermore, whenever multiple nodes are available, programming interface should make it easier to express workload parallelization and distribution. To this end, run-time libraries, middleware and design tools must include innovative approaches to ensure optimal resource scheduling and allocation (see Chapter 6 and Chapter 5).

12.3.1   Reliability

Continuous progresses in miniaturization of electronic chips have a counterpart represented by decreasing reliability of the manufactured devices. Transistor shrinking comes with an increased likelihood of having defective components. This aspect became more evident with latest manufacturing technologies, where more difficulties in achieving acceptable yields have been encountered. By reducing the size of transistors, keeping controls over all the influencing parameters becomes hard. The smaller the size of transistors is, the more likely random fluctuations (e.g., thermal fluctuations) change the features of transistors, causing in some cases their misbehaviour. Still, at such small feature size, transistors exhibit not the same (electrical) characteristics, making even more difficult to create complex systems. Future modular system are expected to largely exploit integration of different technologies, further making more probable erroneous designed system behaviors. To cope with such kind of uncertainty, future system will have to integrate monitoring/correction mechanisms, ranging from complex and smarter embedded testing features to advanced error correcting codes.

With the availability of more (heterogeneous) computing resources, it is expected that the complexity of applications will raise. For instance, the larger availability of massive multi-threading systems (based on multi-/many-cores and accelerators), current programming models (as well as program execution models) become more prone to introduce failures. Indeed, for the programmer it is more difficult to express and correctly map parallelism on such systems. Different approaches for letting compilers extract parallelism will represent one of the future challenges.

12.4 Programming Challenges

On one side, heterogeneity contributes to both improve performance by means of specialization and to keep energy consumption under control. However, on the other side, it poses challenges in the software perspective. Compilers and programming languages have to manage multiple targets and to enable developers to easily express parallelism in their applications. Furthermore, programming languages are demanded to not introduce cumbersome mechanisms at user-level to manage heterogeneity. When target platform is a small IoT device, compilers have to optimize executable code favouring energy efficiency to performance. On the contrary, in large-scale systems, performance may be preferred over energy efficiency. To enable end users to focus on the application logic, abstraction of underlying resources is used. It is expected that, due to the more diverse device availability, such mechanism of abstraction will be further extended. In this respect, the potential return of virtual machines (i.e., abstraction of the processing element instead of the full machine) represents a valuable solution.

Modern programming languages have been designed with a focus on allowing programmers to express functional aspects (or properties of the underlying programmed system) of the problems they tackle, rather than non-functional aspects (i.e., ensure security, request energy efficiency, ensure reliability and trustworthiness, etc.). Historically, such non-functional requirements represented less pressing aspects in the programming phase, which were ensured in later on phases of the development cycle. However, with the ever larger diffusion of digital technologies in many contexts with a corresponding growing of platforms (i.e., the union of hardware and software stack), the situation has been reversed. Nowadays (and in the next future), most of the challenges are in releasing programming languages and platforms allowing the developers to easily express these non-functional properties of the system. Among the others, security (security aspects are discussed in Chapter 10 under the heterogeneity light), reliability, energy efficiency, trustworthiness, and timing and reactivity

(Chapter 11 discussed and analyzed the main aspects related to the support of real-time processing on heterogeneous systems) are the most demanded. Although to some extent, such non-functional properties are already managed, the way of ensuring them still relies on low-level techniques that limit the control by the developers over them, as well as limit the scalability and productivity. Programming languages and frameworks will have to be flexible to serve as many different contexts as possible. To this end, functional and non-functional properties will be ensured by clearly define interfaces among software modules, rather than expressing them explicitly in the program logic. Programming languages inherit mechanisms associated to the programming and execution models they adhere. Current and future heterogeneous systems will pose new challenges which will require to revise programming and execution models to be effectively addressed. Chapter 3 provided an extended overview of the state-of-the-art of modern programming models, highlighting how they tackle the heterogeneous domain. Chapter 4 focused on programming models designed for distributed and parallel (heterogeneous systems) and their evolution.

Programming is also becoming heterogeneous per-se: imperative vs functional, synchronous vs asynchronous, sequential vs concurrent, parallel vs data-flow oriented, homogeneous vs heterogeneous, centralized vs distributed or decentralized, etc. Thus, it is unlikely that modern (future) programming languages will be able to support all such styles and paradigms at once, still preserving the capability of producing consistent, manageable and efficient code. Conversely, it will be more plausible that programming languages and frameworks will be the result of the integration of different, more focused components, each cooperating with the others. The achievement of such kinds of solutions will open the door to more flexible and productive *programming environments*; however, advanced techniques will be required to ensure correctness of the generated code (e.g., methods to ensure correctness-by-construction will be integrated). On the other hand, composable approaches are already fostering the research on *domain-specific languages* (DSL). Compared to general-purpose languages (GPL), DSL should be designed having a single problem domain in mind. Although such claim in some case appears not sharp, DSL gained interest over time. Desired features of DSL include being more declarative than imperative, as well as focusing on expressiveness. There are several ways to implement DSL. Dedicated libraries with clearly defined API can be considered a form of DSL, and can be easily included in a GPL to extend its capabilities. On the other hand, DSL can be implemented as "true" languages. When the former way is chosen, DSL can easily become part of the service offered by the operating systems (through the run-time support or dedicated modules); on the other case, their implementation requires the creation of ad-hoc compiling tools. Chapter 2 discusses deeply required changes in the structure and design of operating systems (OSes) to support modern applications and heterogeneous systems. It also discusses how software modular design approaches can be exploited to make composable the OSes to address new application demands. Modularity can be also at the basis of a more flexible support for DSL. In this case, new challenges arise, since complexity is moved on the compiler side, which becomes responsible for scanning, parsing and semantically checking the input source program. Also, moving from traditional text-based programming environments to graphical-based one will be of help in enabling programmers to focus on expressing "what" rather than "how". As such future challenges, researchers are looking at more "unconventional" approaches to programming. Among the others, the large adoption of artificial intelligence (AI) is seen as a way to enable computing systems to automatically manage different aspects of generating code (ensuring both functional and non-functional properties) [256, 49, 104] from declarative inputs provided by developers (e.g., graphical-based interfaces help to provide input examples to the AI-based program generator).

12.5   Orchestration in Heterogeneous Environments

Starting from the large diffusion of Cloud computing and the development of Software-Defined Networking (SDN) and Network Function Virtualisation (NFV), orchestration became a trending topic in IT research environments. The term orchestration refers to the possibility of a software to configure, allocate, coordinate and manage a set of resources, both hardware and software, in an automated way; the *orchestrator* is a software module in charge of orchestrating a set of resources in order to provide a service. An orchestrator is generally composed of many software modules that concurrently run in a distributed (parallel) infrastructure. Although not explicitly considered part of the orchestrator architecture, middleware plays a key role in managing infrastructural resources. Further, with the large use of heterogeneous devices, middleware became more responsible for exposing the right level of abstraction to the upper levels of the stack, including orchestration modules. Chapter 6 discussed middleware in the context of managing heterogeneous HPC infrastructures and embedded systems.

About 50 billion devices will be connected to the Internet by 2020 [84] and one consequence of this trend is the production of an unprecedented volume of data that must be transmitted, stored and processed. Cloud data centers are one solution for the data processing, since they have enough resources to satisfy the constantly increasing computing request. But the endeavour to use the Cloud as the only solution for this problem involves some other issues: i) Cloud data center resources are centralized, often located in remote places, far from the data generators, and this means that data must be transferred from the generators towards the Cloud; ii) lots of scenarios require rapid reaction to events and this collides with the delay introduced by the data transfer. The Fog Computing concept [185] can be the answer to the problems listed above. Fog computing brings computing power and storage to the edge of the network, as close as possible to the data generators. The cooperation between Fog/Edge nodes and Cloud data centers should be the response to the requirements imposed by the large-scale spread of the IoT devices, but it entails issues related to the correct management of this heterogeneous and distributed resources.

In the last decade, orchestrators constantly improved and now are considered as a response to the growing needs of management and coordination of large amount of heterogeneous and distributed IT resources. Indeed, starting from the management of a limited and centralized set of homogeneous resources, orchestrators improved their interoperability with a great diversity of technologies, such as containerization technologies, network resources, etc. This, combined with the definition of a hierarchy that defines different orchestration layers [98], allows also the possibility to manage resources that are geographically distributed. This is the case of SOAFI [124], one of the major reference architectures for the Fog/Edge computing orchestration, where nodes, each one with a set of limited resources, own a local orchestrator instance that can communicate with the high-level orchestrator. This architecture envisages two layers of orchestration: i) a first layer where Fog orchestrator agents, the orchestrator instances on the edge nodes, manage node resources and communicate status and resource availability data to Fog orchestrator; ii) the high-level orchestrator, a second layer where the Fog orchestrator is responsible for managing, orchestrating, securitying and monitoring of the infrastructure. SOAFI architecture, devoted to Fog computing orchestration, shows that a proper orchestration is essential to ensure service availability, responsiveness and interoperability in a distributed scenario.

Figure 12.4 depicts the architectural role of a modern orchestrator. Modern orchestrators, indeed, are very advanced technologically speaking, being able to provide a good resource management system that can schedule required tasks very quickly following a chosen allocation strategy. Often other functionalities that they include regard the possibility to auto-scale tasks through replication of virtualized resources, the use of load balancers to increase reliability, etc. Despite these advanced features, on the other hand, most orchestrators are very *sectoral*, being developed for a single context of use, being it the Cloud or the Edge computing. These kinds of orchestrators are not suitable to be adopted in complex contexts, where IoT, Edge/Fog and Cloud are used simultaneously. One of the major challenges of orchestration in heterogeneous environments is the development of an orchestrator able to offer similar features as the most advanced ones, but with the possibility to manage heterogeneous contexts, such as the case of resources deployed on the edge and on heterogeneous Clouds. In this way it will be possible to schedule tasks on Cloud or Edge nodes, even with the possibility to move them from one side to the other, depending on some policies defined by the user or administrators. It will be possible, for example, to deploy real-time applications, that require stringent time-processing constraints (based on the latency), allocating

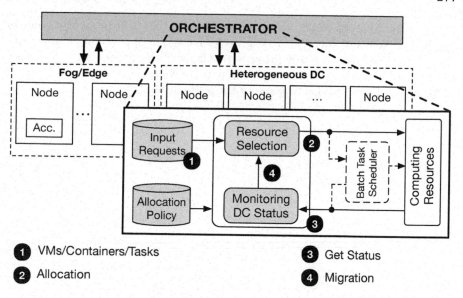

FIGURE 12.4: Orchestration in heterogeneous environments.

required resources on the edge nodes. At the same time, using offloading techniques, it will be possible to move the computations from the Edge to the Cloud and vice versa; e.g., the orchestrator will be able to move active tasks from an under-loaded Edge node to the Cloud, in order to turn off the Edge node thus saving energy (if this action does not violate SLA constraints).

12.6 Conclusion

Digital technologies are transforming our lives and the way we interact with the physical environment. Thanks to modern IoT devices, it is possible to create a digital representation of phenomena that happen in the real world, easing the way we can understand them. The other side of the medal is represented by the massive amount of data that have to be analyzed to extract the information and to have a clear insight of the phenomena. Machine learning and deep learning techniques are revolutionizing the way we approach the analysis of this massive amount of data. In this context *heterogeneity* became the key to efficiently support their execution across infrastructures and platforms that span over the boundaries of single systems and data centers. Future technology evolution will exacerbate such diversification in hardware and software components, also raising new challenges to face. This chapter provided an analysis of the technological trends that will be at the basis of this "heterogeneous future", highlighting the major opportunities and challenges such technologies will bring with them. This chapter aims to create a link with research activities presented in previous chapters, which try to propose possible solutions to some of these challenges.

Bibliography

[1] Amazon accelerated instances. https://aws.amazon.com/ec2/instance-types/p3/.

[2] Amd radeon instinct mi60 gpu. https://www.amd.com/system/files/documents/radeon-instinct-mi60-datasheet.pdf.

[3] Apache OpenWhisk. https://openwhisk.apache.org.

[4] Darknet mnist. https://github.com/ashitani/darknet_mnist.

[5] Ecoscale web-site. http://http://www.ecoscale.eu/.

[6] Fission. https://fission.io.

[7] Fn Project. http://fnproject.io.

[8] Funktion. https://funktion.fabric8.io.

[9] Intel® stratix® 10mx fpga. https://www.intel.it/content/www/it/it/products/programmable/sip/stratix-10-mx.html.

[10] Kubeless. https://kubeless.io.

[11] Making faster. https://leonardoaraujosantos.gitbooks.io/artificial-intelligence/content/making_faster.html.

[12] Microsoft azure instances. https://azure.microsoft.com/en-us/pricing/details/virtual-machines/linux/.

[13] The mnist database of handwritten digits. http://yann.lecun.com/exdb/mnist/.

[14] Nuclio. https://nuclio.io.

[15] Opencl khronos group. https://www.khronos.org/opencl/.

[16] Opencl overview. https://www.khronos.org/opencl/.

[17] OpenFaaS. https://www.openfaas.com.

[18] Project catapult. https://www.microsoft.com/en-us/research/project/project-catapult/.

[19] Quebic. http://quebic-source.github.io.

[20] Quickplay—quickstore by accelize. quickstore.quickplay.io.

[21] Riff. https://projectriff.io.

[22] Serverless. https://serverless.com.

[23] Summit (olcf-4)—supercomputers. https://en.wikichip.org/wiki/supercomputers/summit.

[24] Tiny darknet. https://pjreddie.com/darknet/tiny-darknet/.

[25] ARIANE 5 Flight 501 Failure, Report by the Inquiry Board. http://www-users.math.umn.edu/~arnold/disasters/ariane5rep.html, July 1996.

[26] Bsv high-level hdl, 2018. http://bluespec.com.

[27] Catapult® high-level synthesis, 2018. https://www.mentor.com/hls-lp/catapult-high-level-synthesis/c-systemc-hls.

[28] Intel® high level synthesis compiler, 2018. https://www.intel.com/content/www/us/en/software/programmable/quartus-prime/hls-compiler.html.

[29] Maxcompiler, 2018. https://www.maxeler.com/products/software/maxcompiler/.

[30] Silexica slx tool, 2018. silexica.com.

[31] Vivado high-level synthesis, 2018. https://www.xilinx.com/products/design-tools/vivado/integration/esl-design.html.

[32] S. Kumar A. Postula J. Oberg M. Millberg D. Lindqvist A. Hemani, A. Jantsch. Network on chip: An architecture for billion transistor era. In *Proceeding of the IEEE NorChip Conference*, 2000.

[33] Martın Abadi, Ashish Agarwal, Paul Barham, Eugene Brevdo, Zhifeng Chen, Craig Citro, Greg S Corrado, Andy Davis, Jeffrey Dean, Matthieu Devin, et al. Tensorflow: Large-scale machine learning on heterogeneous distributed systems. *arXiv preprint arXiv:1603.04467*, 2016.

[34] Martín Abadi, Paul Barham, Jianmin Chen, Zhifeng Chen, Andy Davis, Jeffrey Dean, Matthieu Devin, Sanjay Ghemawat, Geoffrey Irving, Michael Isard, et al. Tensorflow: A system for large-scale machine learning. In *12th {USENIX} Symposium on Operating Systems Design and Implementation ({OSDI} 16)*, pages 265–283, 2016.

[35] Martín Abadi, Paul Barham, Jianmin Chen, Zhifeng Chen, Andy Davis, Jeffrey Dean, Matthieu Devin, Sanjay Ghemawat, Geoffrey Irving, Michael Isard, et al. Tensorflow: a system for large-scale machine learning. In *OSDI*, volume 16, pages 265–283, 2016.

[36] Martín Abadi, Mihai Budiu, Úlfar Erlingsson, and Jay Ligatti. Control-Flow Integrity. In *Proceedings of the 12th ACM conference on Computer and Communications Security*, pages 340–353, 2005.

[37] AbsInt Corp. *aiT WCET analyser, http://www.absint.com/ait/*.

[38] Accelerating DNNs with Xilinx Alveo Accelerator Cards. Project Website: https://www.xilinx.com/support/documentation/white_papers/wp504-accel-dnns.pdf.

[39] Mike Accetta, Robert Baron, William Bolosky, David Golub, Richard Rashid, Avadis Tevanian, and Michael Young. Mach: A new kernel foundation for unix development. 1986.

[40] A. Adriahantenaina, H. Charlery, A. Greiner, L. Mortiez, and C.A. Zeferino. Spin: a scalable, packet switched, on-chip micro-network. In *Design, Automation and Test in Europe Conference and Exhibition, 2003*, 2003.

[41] Anant Agarwal, Richard Simoni, John Hennessy, and Mark Horowitz. An evaluation of directory schemes for cache coherence. In *ACM SIGARCH Computer Architecture News*, volume 16, pages 280–298. IEEE Computer Society Press, 1988.

[42] Yuichiro Ajima, Takahiro Kawashima, Takayuki Okamoto, Naoyuki Shida, Kouichi Hirai, Toshiyuki Shimizu, Shinya Hiramoto, Yoshiro Ikeda, Takahide Yoshikawa, Kenji Uchida, et al. The tofu interconnect d. In *2018 IEEE International Conference on Cluster Computing (CLUSTER)*, pages 646–654. IEEE, 2018.

[43] Filipp Akopyan, Jun Sawada, Andrew Cassidy, Rodrigo Alvarez-Icaza, John Arthur, Paul Merolla, Nabil Imam, Yutaka Nakamura, Pallab Datta, Gi-Joon Nam, et al. Truenorth: Design and tool flow of a 65 mw 1 million neuron programmable neurosynaptic chip. *IEEE Transactions on Computer-Aided Design of Integrated Circuits and Systems*, 34(10):1537–1557, 2015.

[44] Rami Al-Rfou, Guillaume Alain, Amjad Almahairi, Christof Angermueller, Dzmitry Bahdanau, Nicolas Ballas, Frédéric Bastien, Justin Bayer, Anatoly Belikov, Alexander Belopolsky, et al. Theano: A python framework for fast computation of mathematical expressions. *arXiv preprint arXiv:1605.02688*, 2016.

[45] Ilkay Altintas, Chad Berkley, Efrat Jaeger, Matthew Jones, Bertram Ludascher, and Steve Mock. Kepler: an extensible system for design and execution of scientific workflows. In *Scientific and Statistical Database Management, 2004. Proceedings. 16th International Conference on*, pages 423–424. IEEE, 2004.

[46] Amazon.com. Announcing Amazon elastic compute cloud (Amazon ec2) - beta. 31 May 2016.

[47] AMD. Aparapi, 2016. `http://aparapi.github.io/`.

[48] Starr Andersen and Vincent Abella. Changes to functionality in microsoft windows xp service pack 2, part 3: Memory protection technologies, Data Execution Prevention. Microsoft TechNet Library, September 2004.

[49] Amir H Ashouri, Gianluca Palermo, John Cavazos, and Cristina Silvano. *Automatic Tuning of Compilers Using Machine Learning*. Springer, 2018.

[50] Cédric Augonnet, Samuel Thibault, Raymond Namyst, and Pierre-André Wacrenier. StarPU: a unified platform for task scheduling on heterogeneous multicore architectures. *Concurrency and Computation: Practice and Experience*, 23(2):187–198, 2011.

[51] Aws.amazon.com. What is cloud computing by Amazon web services — aws. July, 2017.

[52] David F Bacon, Rodric Rabbah, and Sunil Shukla. Fpga programming for the masses. *Communications of the ACM*, 56(4):56–63, 2013.

[53] Rosa M Badia, Javier Conejero, Carlos Diaz, Jorge Ejarque, Daniele Lezzi, Francesc Lordan, Cristian Ramon-Cortes, and Raul Sirvent. Comp superscalar, an interoperable programming framework. *SoftwareX*, 3:32–36, 12 2015.

[54] Rosa M Badia, Javier Conejero, Carlos Diaz, Jorge Ejarque, Daniele Lezzi, Francesc Lordan, Cristian Ramon-Cortes, and Raül Sirvent. Comp superscalar, an interoperable programming framework. *SoftwareX*, 3:32–36, 2015.

[55] Jairo Balart, Alejandro Duran, Marc Gonzàlez, Xavier Martorell, Eduard Ayguadé, and Jesús Labarta. Nanos mercurium: a research compiler for openmp. In *Proceedings of the European Workshop on OpenMP*, volume 8, page 56, 2004.

[56] Barcelona Supercomputing Center. Performance tools. [online], 2016. `http://www.bsc.es/computer-sciences/performance-tools`.

[57] Elena Gabriela Barrantes, David H. Ackley, Stephanie Forrest, and Darko Stefanović. Randomized Instruction Set Emulation. *ACM Transactions on Information and System Security*, 8(1), 2005.

[58] Elena Gabriela Barrantes, David H. Ackley, Trek S. Palmer, Darko Stefanovic, and Dino Dai Zovi. Randomized Instruction Set Emulation to Disrupt Binary Code Injection Attacks. In *ACM Conference on Computer and Communications Security (CCS)*, 2003.

[59] Luiz André Barroso and Urs Hölzle. The case for energy-proportional computing. *Computer*, 40(12):33–37, 2007.

[60] Sanjoy K. Baruah. A general model for recurring real-time tasks. In *RTSS*, 1998.

[61] Sanjoy K. Baruah, Deji Chen, Sergey Gorinsky, and Aloysius K. Mok. Generalized multiframe tasks. *Real-Time Systems*, 17:5–22, 1999.

[62] Andrew Baumann, Paul Barham, Pierre-Evariste Dagand, Tim Harris, Rebecca Isaacs, Simon Peter, Timothy Roscoe, Adrian Schüpbach, and Akhilesh Singhania. The multikernel: A new os architecture for scalable multicore systems. In *Proceedings of the ACM SIGOPS 22Nd Symposium on Operating Systems Principles*, SOSP '09, pages 29–44, New York, NY, USA, 2009. ACM.

[63] Andrew Baumann, Paul Barham, Pierre-Evariste Dagand, Tim Harris, Rebecca Isaacs, Simon Peter, Timothy Roscoe, Adrian Schüpbach, and Akhilesh Singhania. The multikernel: a new os architecture for scalable multicore systems. In *Proceedings of the ACM SIGOPS 22nd symposium on Operating systems principles*, pages 29–44. ACM, 2009.

[64] Tobias Becker, Oskar Mencer, and Georgi Gaydadjiev. Spatial programming with openspl. In *FPGAs for Software Programmers*, pages 81–95. Springer, 2016.

[65] S. Bell and et al. Tile64 - processor: A 64-core soc with mesh interconnect. In *Solid-State Circuits Conference, 2008. ISSCC 2008. Digest of Technical Papers. IEEE International*, 2008.

[66] Anton Beloglazov and Rajkumar Buyya. Energy efficient allocation of virtual machines in cloud data centers. In *Proceedings of the 2010 10th IEEE/ACM International Conference on Cluster, Cloud and Grid Computing*, CCGRID '10, pages 577–578, Washington, DC, USA, 2010. IEEE Computer Society.

[67] Luca Benini and Giovanni De Micheli. Networks on chips: A new soc paradigm. *Computer*, 35(1):70–78, January 2002.

[68] D. Bertozzi and L. Benini. Xpipes: a network-on-chip architecture for gigascale systems-on-chip. *Circuits and Systems Magazine, IEEE*, 2004.

[69] E. Bini, G. Buttazzo, J. Eker, S. Schorr, R. Guerra, G. Fohler, K. Arzen, V. Romero, and C. Scordino. Resource management on multicore systems: The actors approach. *IEEE Micro*, 31(3):72–81, May 2011.

[70] Mark S Birrittella, Mark Debbage, Ram Huggahalli, James Kunz, Tom Lovett, Todd Rimmer, Keith D Underwood, and Robert C Zak. Intel® omni-path architecture: Enabling scalable, high performance fabrics. In *2015 IEEE 23rd Annual Symposium on High-Performance Interconnects*, pages 1–9. IEEE, 2015.

[71] Alan C Bomberger, Norman Hardy, A Peri, Frantz Charles, R Landau, William S Frantz, Jonathan S Shapiro, and Ann C Hardy. The keykos nanokernel architecture. In *In Proc. of the USENIX Workshop on Micro-kernels and Other Kernel Architectures*, 1992.

[72] Jaume Bosch, Xubin Tan, Antonio Filgueras, Miquel Vidal, Marc Mateu, Daniel Jiménez-González, Carlos Álvarez, Xavier Martorell, Eduard Ayguadé, and Jesus Labarta. Application Acceleration on FPGAs with OmpSs@FPGA. In *The 2018 International Conference on Field-Programmable Technology*, December 2018.

[73] Silas Boyd-Wickizer, Haibo Chen, Rong Chen, Yandong Mao, M Frans Kaashoek, Robert Morris, Aleksey Pesterev, Lex Stein, Ming Wu, Yuehua Dai, et al. Corey: An operating system for many cores. In *OSDI*, volume 8, pages 43–57, 2008.

[74] Alexander Branover, Denis Foley, and Maurice Steinman. Amd fusion apu: Llano. *Ieee Micro*, (2):28–37, 2012.

[75] Thomas Brewster. A Massive Number of IoT Cameras Are Hackable – And Now The Next Web Crisis Looms. https://www.forbes.com/sites/thomasbrewster/2017/10/23/reaper-botnet-hacking-iot-cctv-iot-cctv-cameras/, October 2017.

[76] Nathan Brookwood. Amd fusion family of apus: enabling a superior, immersive pc experience. *Insight*, 64(1):1–8, 2010.

[77] François Broquedis, Nathalie Furmento, Brice Goglin, Raymond Namyst, and Pierre-André Wacrenier. Dynamic task and data placement over numa architectures: An openmp runtime perspective. In Matthias S. Müller, Bronis R. de Supinski, and Barbara M. Chapman, editors, *Evolving OpenMP in an Age of Extreme Parallelism*, pages 79–92, Berlin, Heidelberg, 2009. Springer Berlin Heidelberg.

[78] François Broquedis, Thierry Gautier, and Vincent Danjean. libkomp, an efficient openmp runtime system for both fork-join and data flow paradigms. In Barbara M. Chapman, Federico Massaioli, Matthias S. Müller, and Marco Rorro, editors, *OpenMP in a Heterogeneous World*, pages 102–115, Berlin, Heidelberg, 2012. Springer Berlin Heidelberg.

[79] Rainer Buchty, Vincent Heuveline, Wolfgang Karl, and Jan-Philipp Weiss. A survey on hardware-aware and heterogeneous computing on multicore processors and accelerators. *Concurr. Comput. : Pract. Exper.*, 24(7):663–675, May 2012.

[80] Ian Buck, Tim Foley, Daniel Horn, Jeremy Sugerman, Kayvon Fatahalian, Mike Houston, and Pat Hanrahan. Brook for gpus: stream computing on graphics hardware. *ACM transactions on graphics (TOG)*, 23(3):777–786, 2004.

[81] P. Burgio, M. Ruggiero, F. Esposito, M. Marinoni, G. Buttazzo, and L. Benini. Adaptive tdma bus allocation and elastic scheduling: A unified approach for enhancing robustness in multi-core rt systems. In *2010 IEEE International Conference on Computer Design*, pages 187–194, Oct 2010.

[82] G. Buttazzo. *Hard Real-Time Computing Systems: Predictable Scheduling Algorithms and Applications*. Real-Time Systems Series. Springer US, 2004.

[83] Giorgio C. Buttazzo, Giuseppe Lipari, and Luca Abeni. Elastic task model for adaptive rate control. In *RTSS*, 1998.

[84] Charles C. Byers and Patrick Wetterwald. Fog computing distributing data and intelligence for resiliency and scale necessary for iot: The internet of things (ubiquity symposium). *Ubiquity*, 2015(November):4:1–4:12, November 2015.

[85] Stuart Byma, J. Gregory Steffan, Hadi Bannazadeh, Alberto Leon Garcia, and Paul Chow. Fpgas in the cloud: Booting virtualized hardware accelerators with openstack. In *Proceedings of the 2014 IEEE 22Nd International Symposium on Field-Programmable Custom Computing Machines*, FCCM '14, pages 109–116, Washington, DC, USA, 2014. IEEE Computer Society.

[86] Rodrigo N. Calheiros, Rajiv Ranjan, Anton Beloglazov, César A. F. De Rose, and Rajkumar Buyya. Cloudsim: A toolkit for modeling and simulation of cloud computing environments and evaluation of resource provisioning algorithms. *Softw. Pract. Exper.*, 41(1):23–50, January 2011.

[87] Andrew Canis, Jongsok Choi, Mark Aldham, Victor Zhang, Ahmed Kammoona, Jason H. Anderson, Stephen Brown, and Tomasz Czajkowski. Legup: High-level synthesis for fpga-based processor/accelerator systems. In *Proceedings of the 19th ACM/SIGDA International Symposium on Field Programmable Gate Arrays*, FPGA '11, pages 33–36, New York, NY, USA, 2011. ACM.

[88] Andrew Canis and et al. LegUp: An open-source high-level synthesis tool for FPGA-based processor/accelerator systems. *ACM Transactions on Embedded Computing Systems*, 13(2):24:1–24:27, September 2013.

[89] Nicolas Capit, Georges Da Costa, Yiannis Georgiou, Guillaume Huard, Cyrille Martin, Grégory Mounié, Pierre Neyron, and Olivier Richard. A batch scheduler with high level components. In *Cluster Computing and the Grid, 2005. CCGrid 2005. IEEE International Symposium on*, volume 2, pages 776–783. IEEE, 2005.

[90] N. Capodieci, R. Cavicchioli, P. Vogel, A. Marongiu, C. Scordino, and P. Gai. Detailed characterization of platforms. In Deliverable D2.2 of the project HERCULES High-Performance Real-time Architectures for Low-Power Embedded Systems. Technical report, The Hercules Project, 2017.

[91] Nicola Capodieci, Roberto Cavicchioli, Marko Bertogna, and Aingara Paramakuru. Deadline-based scheduling for gpu with preemption support. In *Proceedings of the 39th IEEE Real-Time Systems Symposium*, RTSS '18, pages 119–130. IEEE, 2018.

[92] Nicola Capodieci, Roberto Cavicchioli, Paolo Valente, and Marko Bertogna. Sigamma: Server based integrated gpu arbitration mechanism for memory accesses. In *Proceedings of the 25th International Conference on Real-Time Networks and Systems*, RTNS '17, pages 48–57, New York, NY, USA, 2017. ACM.

[93] Paris Carbone, Asterios Katsifodimos, Stephan Ewen, Volker Markl, Seif Haridi, and Kostas Tzoumas. Apache Flink^{TM}: Stream and Batch Processing in a Single Engine. *IEEE Data Eng. Bull.*, 38:28–38, 2015.

[94] Adrian M. Caulfield, Eric S. Chung, Andrew Putnam, Hari Angepat, Jeremy Fowers, Michael Haselman, Stephen Heil, Matt Humphrey, Puneet Kaur, Joo-Young Kim, Daniel Lo, Todd Massengill, Kalin Ovtcharov, Michael Papamichael, Lisa Woods, Sitaram Lanka, Derek Chiou, and Doug Burger. A cloud-scale acceleration architecture. In *The 49th Annual IEEE/ACM International Symposium on Microarchitecture*, MICRO-49, pages 7:1–7:13, Piscataway, NJ, USA, 2016. IEEE Press.

[95] R. Cavicchioli, N. Capodieci, and M. Bertogna. Memory interference characterization between cpu cores and integrated gpus in mixed-criticality platforms. In *2017 22nd IEEE International Conference on Emerging Technologies and Factory Automation (ETFA)*, pages 1–10, Sept 2017.

[96] Lucien M Censier and Paul Feautrier. A new solution to coherence problems in multicache systems. *Computers, IEEE Transactions on*, 100(12):1112–1118, 1978.

[97] Goolge Press Center. Cloud term, 2006. Search Engine Strategies Conference.

[98] Ivano Cerrato, Alex Palesandro, Fulvio Risso, Marc Suñ, Vinicio Vercellone, and Hagen Woesner. Toward dynamic virtualized network services in telecom operator networks. *Computer Networks*, 92:380 – 395, 2015. Software Defined Networks and Virtualization.

[99] Anantha P Chandrakasan, Miodrag Potkonjak, Renu Mehra, Jan Rabaey, and Robert W Broderse. Optimizing power using transformations. *Computer-Aided Design of Integrated Circuits and Systems, IEEE Transactions on*, 14(1):12–31, 1995.

[100] Robert Charette. This car runs on code, feb 2009.

[101] Philippe Charles, Christian Grothoff, Vijay Saraswat, Christopher Donawa, Allan Kielstra, Kemal Ebcioglu, Christoph Von Praun, and Vivek Sarkar. X10: an object-oriented approach to non-uniform cluster computing. *ACM Sigplan Notices*, 40(10):519–538, 2005.

[102] C. Chen, K. Li, A. Ouyang, Z. Tang, and K. Li. Gpu-accelerated parallel hierarchical extreme learning machine on flink for big data. *IEEE Transactions on Systems, Man, and Cybernetics: Systems*, 47(10):2740–2753, Oct 2017.

[103] C. Chen, K. Li, A. Ouyang, Z. Zeng, and K. Li. GFlink: An In-Memory Computing Architecture on Heterogeneous CPU-GPU Clusters for Big Data. *IEEE Transactions on Parallel and Distributed Systems*, 29(6):1275–1288, June 2018.

[104] Tianqi Chen, Lianmin Zheng, Eddie Yan, Ziheng Jiang, Thierry Moreau, Luis Ceze, Carlos Guestrin, and Arvind Krishnamurthy. Learning to optimize tensor programs. In *Advances in Neural Information Processing Systems*, pages 3393–3404, 2018.

[105] Sharan Chetlur, Cliff Woolley, Philippe Vandermersch, Jonathan Cohen, John Tran, Bryan Catanzaro, and Evan Shelhamer. cudnn: Efficient primitives for deep learning. *CoRR*, abs/1410.0759, 2014.

[106] Minsik Cho and Daniel Brand. Mec: memory-efficient convolution for deep neural network. *arXiv preprint arXiv:1706.06873*, 2017.

[107] Jack Choquette, Olivier Giroux, and Denis Foley. Volta: performance and programmability. *IEEE Micro*, 38(2):42–52, 2018.

[108] Nick Christoulakis, George Christou, Elias Athanasopoulos, and Sotiris Ioannidis. HCFI: Hardware-enforced Control-Flow Integrity. In *Proceedings of the Sixth ACM Conference on Data and Application Security and Privacy*, CODASPY '16, pages 38–49, New York, NY, USA, 2016. ACM.

[109] J.-P. Colinge. Multigate transistors: Pushing moore's law to the limit. In *Simulation of Semiconductor Processes and Devices (SISPAD), 2014 International Conference on*, pages 313–316, Sept 2014.

[110] Ronan Collobert, Samy Bengio, and Johnny Mariéthoz. Torch: a modular machine learning software library. Technical report, Technical Report IDIAP-RR 02-46, IDIAP, 2002.

[111] Jason Cong, Zhenman Fang, Michael Lo, Hanrui Wang, Jingxian Xu, and Shaochong Zhang. Understanding performance differences of fpgas and gpus: (abtract only). In *Proceedings of the 2018 ACM/SIGDA International Symposium on Field-Programmable Gate Arrays*, FPGA '18, pages 288–288, New York, NY, USA, 2018. ACM.

[112] UPC Consortium et al. Upc language specifications v1. 2. *Lawrence Berkeley National Laboratory*, 2005.

[113] Intel Corporation. Intel® xeon scalable platform brief.

[114] NVIDIA Corporation. Nvidia turing architecture whitepaper.

[115] Corinna Cortes and Vladimir Vapnik. Support-vector networks. *Machine learning*, 20(3):273–297, 1995.

[116] Thomas Cover and Peter Hart. Nearest neighbor pattern classification. *IEEE transactions on information theory*, 13(1):21–27, 1967.

[117] S.P. Crago and J.P. Walters. Heterogeneous cloud computing: The way forward. *Computer*, 48(1):59–61, 2015.

[118] Nvidia CUDA, 2007. http://developer.nvidia.com/.

[119] Chris Cummins, Pavlos Petoumenos, Zheng Wang, and Hugh Leather. End-to-end deep learning of optimization heuristics. In *Parallel Architectures and Compilation Techniques (PACT), 2017 26th International Conference on*, pages 219–232. IEEE, 2017.

[120] Leonardo Dagum and Ramesh Menon. Openmp: An industry-standard api for shared-memory programming. *IEEE Comput. Sci. Eng.*, 5(1):46–55, January 1998.

[121] William J. Dally and Brian Towles. Route packets, not wires: On-chip inteconnection networks. In *Proceedings of the 38th Annual Design Automation Conference*, DAC '01, pages 684–689, New York, NY, USA, 2001. ACM.

[122] Mike Davies, Narayan Srinivasa, Tsung-Han Lin, Gautham Chinya, Yongqiang Cao, Sri Harsha Choday, Georgios Dimou, Prasad Joshi, Nabil Imam, Shweta Jain, et al. Loihi: a neuromorphic manycore processor with on-chip learning. *IEEE Micro*, 38(1):82–99, 2018.

[123] Miyuru Dayarathna, Yonggang Wen, and Rui Fan. Data center energy consumption modeling: A survey. *IEEE Communications Surveys & Tutorials*, 18:732–794, 2016.

[124] M. S. de Brito, S. Hoque, T. Magedanz, R. Steinke, A. Willner, D. Nehls, O. Keils, and F. Schreiner. A service orchestration architecture for fog-enabled infrastructures. In *2017 Second International Conference on Fog and Mobile Edge Computing (FMEC)*, pages 127–132, May 2017.

[125] Benoît Dupont de Dinechin, Pierre Guironnet de Massas, Guillaume Lager, Clément Léger, Benjamin Orgogozo, Jérôme Reybert, and Thierry Strudel. A distributed run-time environment for the kalray mppa®-256 integrated manycore processor. *Procedia Computer Science*, 18:1654–1663, 2013.

[126] de Lemos, Rogério, et. al. *Software Engineering for Self-Adaptive Systems: A Second Research Roadmap*, pages 1–32. Springer Berlin Heidelberg, Berlin, Heidelberg, 2013.

[127] Jeffrey Dean and Sanjay Ghemawat. Mapreduce: simplified data processing on large clusters. *Communications of the ACM*, 51(1):107–113, 2008.

[128] R. DiCecco, G. Lacey, J. Vasiljevic, P. Chow, G. Taylor, and S. Areibi. Caffeinated fpgas: Fpga framework for convolutional neural networks. In *2016 International Conference on Field-Programmable Technology (FPT)*, pages 265–268, Dec 2016.

[129] Shifei Ding, Han Zhao, Yanan Zhang, Xinzheng Xu, and Ru Nie. Extreme learning machine: Algorithm, theory and applications. *Artif. Intell. Rev.*, 44(1):103–115, June 2015.

[130] K Djemame, R Bosch, R Kavanagh, P Alvarez, J Ejarque, J Guitart, and L Blasi. PaaS-IaaS Inter-Layer Adaptation in an Energy-Aware Cloud Environment. *IEEE Transactions on Sustainable Computing*, 2(2):127–139, apr 2017.

[131] K Djemame and V Kelefouras. A methodology for efficient code optimizations and memory management, March 2018.

[132] Samuel Dodge and Lina Karam. A study and comparison of human and deep learning recognition performance under visual distortions. 05 2017.

[133] Katerina Doka, Nikolaos Papailiou, Victor Giannakouris, Dimitrios Tsoumakos, and Nectarios Koziris. Mix 'n' match multi-engine analytics. In *Big Data (Big Data), 2016 IEEE International Conference on*, pages 194–203. IEEE, 2016.

[134] R. Dolbeau, S. Bihan, and F. Bodin. HMPP: A hybrid multi-core parallel programming environment. In *First Workshop on General Purpose Processing on Graphics Processing Units*, October 2007.

[135] Jack Dongarra. Report on the sunway taihulight system. Technical report, University of Tennessee, Department of Electrical Engineering and Computer Science, 2016.

[136] Alejandro Duran, Eduard Ayguadé, Rosa M. Badia, Jesús Labarta, Luis Martinell, Xavier Martorell, and Judit Planas. Ompss: a Proposal for Programming Heterogeneous Multi-Core Architectures. *Parallel Processing Letters*, 21(2):173–193, 2011.

[137] Alejandro Duran et al. Ompss: a proposal for programming heterogeneous multi-core architectures. *Parallel Processing Letters*, 21(02):173–193, 2011.

[138] Marc Duranton, Koen De Bosschere, Bart Coppens, Christian Gamrat, Madeleine Gray, Harm Munk, Emre Ozer, Tullio Vardanega, and Olivier Zendra. The hipeac vision 2019, 2019.

[139] G. Durelli, M. Coppola, K. Djafarian, G. Kornaros, A. Miele, M. Paolino, Oliver Pell, Christian Plessl, M. D. Santambrogio, and C. Bolchini. Save: Towards efficient resource management in heterogeneous system architectures. In Diana Goehringer, Marco Domenico Santambrogio, João M. P. Cardoso, and Koen Bertels, editors, *Reconfigurable Computing: Architectures, Tools, and Applications*, pages 337–344, Cham, 2014. Springer International Publishing.

[140] P F Dutot, Y Georgiou, D Glesser, L Lefevre, M Poquet, and I Rais. Towards Energy Budget Control in HPC. In *2017 17th IEEE/ACM International Symposium on Cluster, Cloud and Grid Computing (CCGRID)*, pages 381–390, 2017.

[141] eclipse foundation. The platform for open innovation and collaboration, 2018. http://www.eclipse.org/.

[142] Ecoscale Consortium. Project description. [online], 2018. http://ecoscale.eu/project-description.html.

[143] L. Eisen. Ibm power6 accelerators: Vmx and dfu, 2007.

[144] D. R. Engler, M. F. Kaashoek, and J. O'Toole, Jr. Exokernel: An operating system architecture for application-level resource management. In *Proceedings of the Fifteenth ACM Symposium on Operating Systems Principles*, SOSP '95, pages 251–266, New York, NY, USA, 1995. ACM.

[145] Hadi Esmaeilzadeh, Emily Blem, Renee St. Amant, Karthikeyan Sankaralingam, and Doug Burger. Dark silicon and the end of multicore scaling. In *Proceedings of the 38th Annual International Symposium on Computer Architecture*, ISCA '11, pages 365–376, New York, NY, USA, 2011. ACM.

[146] Hadi Esmaeilzadeh, Emily Blem, Renee St. Amant, Karthikeyan Sankaralingam, and Doug Burger. Dark silicon and the end of multicore scaling. *SIGARCH Comput. Archit. News*, 39(3):365–376, June 2011.

[147] Jack Dongarra et al. The international exascale software project roadmap. *The International Journal of High Performance Computing Applications*, 25(1):3–60, 2011.

[148] Evidence srl. ERIKA Enterprise kernel.

[149] Xiaobo Fan, Wolf-Dietrich Weber, and Luiz Andre Barroso. Power provisioning for a warehouse-sized computer. In *ACM SIGARCH Computer Architecture News*, volume 35, pages 13–23. ACM, 2007.

[150] Christos K. Filelis-Papadopoulos, George A. Gravvanis, and Panagiotis E. Kyziropoulos. A framework for simulating large scale cloud infrastructures. *Future Gener. Comput. Syst.*, 79(P2):703–714, February 2018.

[151] M. Filho, Oliveira, C.C.M. Monteiro, P.R.M. Inácio, and M. Freire. Cloudsim plus: a cloud computing simulation framework pursuing software engineering principles for improved modularity, extensibility and correctness. In *IFIP/IEEE International Symp. on Integrated Network Management - IFIP/IEEE IM*, 2017.

[152] Nadeem Firasta and et al. Intel avx: New frontiers in performance improvements and energy efficiency – white paper, 2008.

[153] Daniel Firestone, Andrew Putnam, Sambhrama Mundkur, Derek Chiou, Alireza Dabagh, Mike Andrewartha, Hari Angepat, Vivek Bhanu, Adrian Caulfield, Eric Chung, et al. Azure accelerated networking: Smartnics in the public cloud. In *15th {USENIX} Symposium on Networked Systems Design and Implementation ({NSDI} 18)*, pages 51–66, 2018.

[154] Ronald A Fisher. The use of multiple measurements in taxonomic problems. *Annals of eugenics*, 7(2):179–188, 1936.

[155] J Flich, G Agosta, P Ampletzer, D A Alonso, C Brandolese, E Cappe, A Cilardo, L Dragić, A Dray, A Duspara, W Fornaciari, G Guillaume, Y Hoornenborg, A Iranfar, M Kovač, S Libutti, B Maitre, J M Martínez, G Massari, H Mlinarić, E Papastefanakis, T Picornell, I Piljić, A Pupykina, F Reghenzani, I Staub, R Tornero, M Zapater, and D Zoni. MANGO: Exploring Manycore Architectures for Next-GeneratiOn HPC Systems. In *2017 Euromicro Conference on Digital System Design (DSD)*, pages 478–485, 2017.

[156] International Organization for Standardization / Technical Committee 22 (ISO/TC 22). ISO/DIS 26262-1 - Road vehicles—Functional safety. Technical report, International Organization for Standardization / Technical Committee 22 (ISO/TC 22), Geneva, Switzerland, July 2009.

[157] B. Fort, A. Canis, J. Choi, N. Calagar, R. Lian, S. Hadjis, Y. T. Chen, M. Hall, B. Syrowik, T. Czajkowski, S. Brown, and J. Anderson. Automating the Design of Processor/Accelerator Embedded Systems with LegUp High-Level Synthesis. In *2014 12th IEEE International Conference on Embedded and Ubiquitous Computing*, pages 120–129, Aug 2014.

[158] HSA Foundation. HSA Platform System Architecture Specification 1.1, 2017.

[159] HSA Foundation. HSA Programmer's Reference Manual: HSAIL Virtual ISA and Programming Model, 2017.

[160] HSA Foundation. ROCm, a new Era in Open GPU Computing, 2017. https://radeonopencompute.github.io/index.html.

[161] Jeremy Fowers, Kalin Ovtcharov, Michael Papamichael, Todd Massengill, Ming Liu, Daniel Lo, Shlomi Alkalay, Michael Haselman, Logan Adams, Mahdi Ghandi, et al. A configurable cloud-scale dnn processor for real-time ai. In *Proceedings of the 45th Annual International Symposium on Computer Architecture*, pages 1–14. IEEE Press, 2018.

[162] Juan Fumero. *Accelerating Interpreted Programming Languages on GPUs with Just-In-Time and Runtime Optimisations.* PhD thesis, The University of Edinburgh, UK, August 2017.

[163] Steve B Furber, David R Lester, Luis A Plana, Jim D Garside, Eustace Painkras, Steve Temple, and Andrew D Brown. Overview of the spinnaker system architecture. *IEEE Transactions on Computers*, 62(12):2454–2467, 2013.

[164] Paolo Gai, Enrico Bini, Giuseppe Lipari, Marco Di Natale, and Luca Abeni. Architecture for a portable open source real time kernel environment. In *In Proceedings of the Second Real-Time Linux Workshop and Hand's on Real-Time Linux Tutorial*, 2000.

[165] Brian Gaide, Dinesh Gaitonde, Chirag Ravishankar, and Trevor Bauer. Xilinx adaptive compute acceleration platform: Versal tm architecture. In *Proceedings of the 2019 ACM/SIGDA International Symposium on Field-Programmable Gate Arrays*, pages 84–93. ACM, 2019.

[166] Daniel D Gajski, Nikil D Dutt, Allen CH Wu, and Steve YL Lin. *High—Level Synthesis: Introduction to Chip and System Design.* Springer Science & Business Media, 2012.

[167] Anshul Gandhi, Yuan Chen, Daniel Gmach, Martin F. Arlitt, and Manish Marwah. Minimizing data center SLA violations and power consumption via hybrid resource provisioning. In *2011 International Green Computing Conference and Workshops, IGCC 2012, Orlando, FL, USA, July 25-28, 2011*, pages 1–8, 2011.

[168] Jaspreet Gandhi, Boon Ang, Tom Lee, Henley Liu, Myongseob Kim, Ho Hyung Lee, Gamal Refai-Ahmed, Hong Shi, and Suresh Ramalingam. 2.5 d fpga-hbm integration challenges. In *International Symposium on Microelectronics*, volume 2017, pages 000336–000341. International Microelectronics Assembly and Packaging Society, 2017.

[169] Steven Gay, Renaud Hartert, Christophe Lecoutre, and Pierre Schaus. Conflict ordering search for scheduling problems. In Gilles Pesant, editor, *Principles and Practice of Constraint Programming*, pages 140–148, Cham, 2015. Springer International Publishing.

[170] Robert Geirhos, David H. J. Janssen, Heiko H. Schütt, Jonas Rauber, Matthias Bethge, and Felix A. Wichmann. Comparing deep neural networks against humans: object recognition when the signal gets weaker. *CoRR*, abs/1706.06969, 2017.

[171] Konstantinos Georgopoulos, Iakovos Mavroidis, Luciano Lavagno, Ioannis Papaefstathiou, and Konstantin Bakanov. *Energy-Efficient Heterogeneous Computing at exaSCALE—ECOSCALE*, pages 199–213. Springer International Publishing, Cham, 2019.

[172] Balazs Gerofi, Masamichi Takagi, Atsushi Hori, Gou Nakamura, Tomoki Shirasawa, and Yutaka Ishikawa. On the scalability, performance isolation and device driver transparency of the ihk/mckernel hybrid lightweight kernel. In *Parallel and Distributed Processing Symposium, 2016 IEEE International*, pages 1041–1050. IEEE, 2016.

[173] Saugata Ghose, Kevin Hsieh, Amirali Boroumand, Rachata Ausavarungnirun, and Onur Mutlu. Enabling the adoption of processing-in-memory: Challenges, mechanisms, future research directions. *arXiv preprint arXiv:1802.00320*, 2018.

[174] Jeremy Goecks, Anton Nekrutenko, and James Taylor. Galaxy: a comprehensive approach for supporting accessible, reproducible, and transparent computational research in the life sciences. *Genome biology*, 11(8):R86, 2010.

[175] Klodiana Goga, Antonio Parodi, Pietro Ruiu, and Olivier Terzo. Performance analysis of wrf simulations in a public cloud and hpc environment. In *Conference on Complex, Intelligent, and Software Intensive Systems*, pages 384–396. Springer, 2017.

[176] Klodiana Goga, Luca Pilosu, Antonio Parodi, Martina Lagasio, and Olivier Terzo. Performance of wrf cloud resolving simulations with data assimilation on public cloud and hpc environments. In *Conference on Complex, Intelligent, and Software Intensive Systems*, pages 161–171. Springer, 2018.

[177] Jiong Gong, Haihao Shen, Guoming Zhang, Xiaoli Liu, Shane Li, Ge Jin, Niharika Maheshwari, Evarist Fomenko, and Eden Segal. Highly efficient 8-bit low precision inference of convolutional neural networks with intelcaffe. In *ReQuEST@ASPLOS*, 2018.

[178] James R Goodman. Using cache memory to reduce processor-memory traffic. In *ACM SIGARCH Computer Architecture News*, volume 11, pages 124–131. ACM, 1983.

[179] Sven Goossens, Karthik Chandrasekar, Benny Akesson, and Kees Goossens. *Memory Controllers for Mixed-Time-Criticality Systems: Architectures, Methodologies and Trade-Offs*. Springer Publishing Company, Incorporated, 2016.

[180] Albert Greenberg, James Hamilton, David A Maltz, and Parveen Patel. The cost of a cloud: research problems in data center networks. *ACM SIGCOMM computer communication review*, 39(1):68–73, 2008.

[181] Dominik Grewe and Michael F. P. O'Boyle. A static task partitioning approach for heterogeneous systems using opencl. In Jens Knoop, editor, *Compiler Construction*, pages 286–305, Berlin, Heidelberg, 2011. Springer Berlin Heidelberg.

[182] William D Gropp, William Gropp, Ewing Lusk, Anthony Skjellum, and Argonne Distinguished Fellow Emeritus Ewing Lusk. *Using MPI: portable parallel programming with the message-passing interface*, volume 1. MIT press, 1999.

[183] T. Grosser and T. Hoefler. Polly-ACC: Transparent compilation to heterogeneous hardware. In *Proceedings of the the 30th International Conference on Supercomputing (ICS'16)*, Jun. 2016.

[184] Max Grossman and Vivek Sarkar. Swat: A programmable, in-memory, distributed, high-performance computing platform. In *HPDC*, 2016.

[185] Jayavardhana Gubbi, Rajkumar Buyya, Slaven Marusic, and Marimuthu Palaniswami. Internet of things (iot): A vision, architectural elements, and future directions. *Future Gener. Comput. Syst.*, 29(7):1645–1660, September 2013.

[186] S. Hariri H. Topcuouglu and M. y. Wu. Performance-Effective and Low-Complexity Task Scheduling for Heterogeneous Computing. In *IEEE Trans. Parallel Distrib. Syst., vol. 13, no. 3*, 2002.

[187] Panagiotis E Hadjidoukas and Vassilios V Dimakopoulos. Nested parallelism in the ompi openmp/c compiler. In *European Conference on Parallel Processing*, pages 662–671. Springer, 2007.

[188] Mohammad Hamad, Zain A. H. Hammadeh, Selma Saidi, Vassilis Prevelakis, and Rolf Ernst. Prediction of abnormal temporal behavior in real-time systems. In *Proceedings of the 33rd Annual ACM Symposium on Applied Computing*, SAC '18, pages 359–367. ACM, 2018.

[189] Mohammad Hamad, Marcus Nolte, and Vassilis Prevelakis. Towards comprehensive threat modeling for vehicles. In *1st Workshop on Security and Dependability of Critical Embedded Real-Time Systems*, 2016.

[190] Mohammad Hamad, Marcus Nolte, and Vassilis Prevelakis. A framework for policy based secure intra vehicle communication. In *2017 IEEE Vehicular Networking Conference (VNC)*, pages 1–8, Nov 2017.

[191] Mohammad Hamad and Vassilis Prevelakis. Implementation and performance evaluation of embedded ipsec in microkernel os. In *Computer Networks and Information Security (WSCNIS), 2015 World Symposium on*, pages 1–7. IEEE, 2015.

[192] Mohammad Hamad, Marinos Tsantekidis, and Vassilis Prevelakis. Redzone: Towards an intrusion response framework for intra-vehicle system.

In *the 5th International Conference on Vehicle Technology and Intelligent Transport Systems (VEHITS)*, Crete, Greece, May 2019.

[193] Paul Harvey, Konstantin Bakanov, Ivor Spence, and Dimitrios S. Nikolopoulos. A scalable runtime for the ecoscale heterogeneous exascale hardware platform. In *Proceedings of the 6th International Workshop on Runtime and Operating Systems for Supercomputers*, ROSS '16, pages 7:1–7:8, New York, NY, USA, 2016. ACM.

[194] Paul Harvey, Konstantin Bakanov, Ivor Spence, and Dimitrios S. Nikolopoulos. A scalable runtime for the ecoscale heterogeneous exascale hardware platform. In *Proceedings of the 6th International Workshop on Runtime and Operating Systems for Supercomputers*, ROSS '16, pages 7:1–7:8, New York, NY, USA, 2016. ACM.

[195] Akihiro Hayashi, Kazuaki Ishizaki, Gita Koblents, and Vivek Sarkar. Machine-learning-based performance heuristics for runtime cpu/gpu selection. In *PPPJ*, 2015.

[196] Robert L Henderson. Job scheduling under the portable batch system. In *Workshop on Job Scheduling Strategies for Parallel Processing*, pages 279–294. Springer, 1995.

[197] Catherine F. Higham and Desmond J. Higham. Deep learning: An introduction for applied mathematicians. January 2018.

[198] Benjamin Hindman, Andy Konwinski, Matei Zaharia, Ali Ghodsi, Anthony D. Joseph, Randy Katz, Scott Shenker, and Ion Stoica. Mesos: A platform for fine-grained resource sharing in the data center. In *Proceedings of the 8th USENIX Conference on Networked Systems Design and Implementation*, NSDI'11, pages 295–308, Berkeley, CA, USA, 2011. USENIX Association.

[199] Zhe-Mao Hsu, I-Yao Chuang, Wen-Chien Su, Jen-Chieh Yeh, Jen-Kuei Yang, and Shau-Yin Tseng. System performance analyses on pac duo esl virtual platform. In *Intelligent Information Hiding and Multimedia Signal Processing, 2009. IIH-MSP'09. Fifth International Conference on*, pages 406–409. IEEE, 2009.

[200] Wei Hu, Jason Hiser, Dan Williams, Adrian Filipi, Jack W. Davidson, David Evans, John C. Knight, Anh Nguyen-Tuong, and Jonathan Rowanhill. Secure and Practical Defense Against Code-Injection Attacks using Software Dynamic Translation. In *ACM SIGPLAN/SIGOPS Conference on Virtual Execution Environments (VEE)*, 2006.

[201] I. Mavroidis, et. al. Ecoscale: Reconfigurable computing and runtime system for future exascale systems. In *2016 Design, Automation Test in Europe Conference Exhibition (DATE)*, pages 696–701, March 2016.

[202] Forrest N Iandola, Song Han, Matthew W Moskewicz, Khalid Ashraf, William J Dally, and Kurt Keutzer. Squeezenet: Alexnet-level accuracy with 50x fewer parameters and 0.5 mb model size. *arXiv preprint arXiv:1602.07360*, 2016.

[203] Forrest N. Iandola, Matthew W. Moskewicz, Khalid Ashraf, Song Han, William J. Dally, and Kurt Keutzer. Squeezenet: Alexnet-level accuracy with 50x fewer parameters and ¡1mb model size. *CoRR*, abs/1602.07360, 2016.

[204] Didac Gil De La Iglesia and Danny Weyns. Mape-k formal templates to rigorously design behaviors for self-adaptive systems. *ACM Transactions on Autonomous Adaptive Systems*, 10(3):15:1–15:31, September 2015.

[205] Ross Ihaka and Robert Gentleman. R: A language for data analysis and graphics. *Journal of Computational and Graphical Statistics*, 5(3):299–314, 1996.

[206] Adapteva Inc. Epiphany architecture reference.

[207] Infineon Technologies AG. Tricore AURIX Family.

[208] Intel. Intel fpga sdk for opencl. 2017.

[209] Intel Corp. Quartus Prime, September 2017.

[210] ITRS. International technology roadmap for semiconductors–system drivers, 2013. http://www.itrs.net.

[211] Dana A Jacobsen, Julien C Thibault, and Inanc Senocak. An mpi-cuda implementation for massively parallel incompressible flow computations on multi-gpu clusters. In *48th AIAA aerospace sciences meeting and exhibit*, volume 16, 2010.

[212] Grant A Jacoby and NathanielJ Davis. Battery-based intrusion detection. In *Global Telecommunications Conference, 2004. GLOBECOM'04. IEEE*, volume 4, pages 2250–2255. IEEE.

[213] Joe Jeddeloh and Brent Keeth. Hybrid memory cube new dram architecture increases density and performance. In *2012 symposium on VLSI technology (VLSIT)*, pages 87–88. IEEE, 2012.

[214] Yangqing Jia, Evan Shelhamer, Jeff Donahue, Sergey Karayev, Jonathan Long, Ross Girshick, Sergio Guadarrama, and Trevor Darrell. Caffe: Convolutional architecture for fast feature embedding. In *Proceedings of the 22Nd ACM International Conference on Multimedia*, MM '14, pages 675–678, New York, NY, USA, 2014. ACM.

[215] Yangqing Jia, Evan Shelhamer, Jeff Donahue, Sergey Karayev, Jonathan Long, Ross Girshick, Sergio Guadarrama, and Trevor Darrell. Caffe: Convolutional architecture for fast feature embedding. In *Proceedings of the 22nd ACM international conference on Multimedia*, pages 675–678. ACM, 2014.

[216] Zhe Jia, Marco Maggioni, Benjamin Staiger, and Daniele P Scarpazza. Dissecting the nvidia volta gpu architecture via microbenchmarking. *arXiv preprint arXiv:1804.06826*, 2018.

[217] Java bindings for OpenCL, 2017. http://www.jocl.org/.

[218] K Johnson, R Sinha, R Calinescu, and J Ruan. A Multi-agent Framework for Dependable Adaptation of Evolving System Architectures. In *2015 41st Euromicro Conference on Software Engineering and Advanced Applications*, pages 159–166, aug 2015.

[219] Jithin Jose, Sreeram Potluri, Hari Subramoni, Xiaoyi Lu, Khaled Hamidouche, Karl Schulz, Hari Sundar, and Dhabaleswar K Panda. Designing scalable out-of-core sorting with hybrid mpi+ pgas programming models. In *Proceedings of the 8th International Conference on Partitioned Global Address Space Programming Models*, page 7. ACM, 2014.

[220] Norman P Jouppi, Cliff Young, Nishant Patil, David Patterson, Gaurav Agrawal, Raminder Bajwa, Sarah Bates, Suresh Bhatia, Nan Boden, Al Borchers, et al. In-datacenter performance analysis of a tensor processing unit. In *Computer Architecture (ISCA), 2017 ACM/IEEE 44th Annual International Symposium on*, pages 1–12. IEEE, 2017.

[221] K. Djemame, et. al. Tango: Transparent heterogeneous hardware architecture deployment for energy gain in operation. In S. Tamarit, G. Vigueras, M. Carro, and J. Marino, editors, *Proceedings of the First Workshop on Program Transformation for Programmability in Heterogeneous Architectures*, Barcelona, Spain, March 2016.

[222] Christoforos Kachris, Georgios Sirakoulis, and Dimitrios Soudris. Network function virtualization based on fpgas: A framework for all-programmable network devices. *arXiv preprint arXiv:1406.0309*, 2014.

[223] David R Kaeli, Perhaad Mistry, Dana Schaa, and Dong Ping Zhang. *Heterogeneous computing with OpenCL 2.0*. Morgan Kaufmann, 2015.

[224] Kalray Corporation. Many-core Kalray MPPA, 2012.

[225] Krishna Kant. Data center evolution: A tutorial on state of the art, issues, and challenges. *Computer Networks*, 53(17):2939–2965, 2009.

[226] Faraydon Karim, Anh Nguyen, and Sujit Dey. An interconnect architecture for networking systems on chips. *IEEE Micro*, 22(5), 2002.

[227] Michael Kaufmann and Kornilios Kourtis. The hcl scheduler: Going all-in on heterogeneity. In *9th USENIX Workshop on Hot Topics in Cloud Computing (HotCloud 17)*, Santa Clara, CA, 2017. USENIX Association.

[228] J. Kavalieros and et al. Tri-gate transistor architecture with high-k gate dielectrics, metal gates and strain engineering. In *VLSI Technology, 2006. Digest of Technical Papers. 2006 Symposium on*, pages 50–51, 2006.

[229] Gaurav S. Kc, Angelos D. Keromytis, and Vassilis Prevelakis. Countering Code-injection Attacks with Instruction-set Randomization. In *Proceedings of the 10th ACM Conference on Computer and Communications Security*, CCS '03, pages 272–280, New York, NY, USA, 2003. ACM.

[230] Simon C. Steely Jr. Jinjie Tang Alan G. Gara Kermin E. Fleming JR., Kent D. Glossop. Processors, methods, and systems with a configurable spatial accelerator, 2018. US Patent No. US20180189231A1.

[231] Khronos. SPIR-V Specification Provisional, Version 1.1, 2017. https://www.khronos.org/registry/spir-v/specs/1.1/SPIRV.pdf.

[232] John Kim, James Balfour, and William Dally. Flattened butterfly topology for on-chip networks. In *Proceedings of the 40th Annual IEEE/ACM International Symposium on Microarchitecture*, MICRO 40, pages 172–182. IEEE Computer Society, 2007.

[233] Joonyoung Kim and Younsu Kim. Hbm: Memory solution for bandwidth-hungry processors. In *2014 IEEE Hot Chips 26 Symposium (HCS)*, pages 1–24. IEEE, 2014.

[234] V Klos, T Gothel, and S Glesner. Adaptive Knowledge Bases in Self-Adaptive System Design. In *2015 41st Euromicro Conference on Software Engineering and Advanced Applications*, pages 472–478, aug 2015.

[235] Fanxin Kong and Xue Liu. A Survey on Green-Energy-Aware Power Management for Datacenters. *ACM Computing Surveys*, 47(2):1–38, nov 2014.

[236] Angeliki Kritikakou, Francky Catthoor, Vasilios I. Kelefouras, and Costas E. Goutis. A systematic approach to classify design-time global scheduling techniques. *ACM Comput. Surv.*, 45(2):14:1–14:30, 2013.

[237] Christian Krupitzer, Felix Maximilian Roth, Sebastian VanSyckel, Gregor Schiele, and Christian Becker. A survey on engineering approaches for self-adaptive systems. *Pervasive and Mobile Computing*, 17:184 – 206, 2015.

[238] Mohit Kumar. Chinese Spying Chips Found Hidden On Servers Used By US Companies. https://thehackernews.com/2018/10/china-spying-server-chips.html, October 2018.

[239] Rakesh Kumar, Victor Zyuban, and Dean M. Tullsen. Interconnections in multi-core architectures: Understanding mechanisms, overheads and scaling. *SIGARCH Comput. Archit. News*, 33(2), 2005.

[240] Adam Lackorzynski and Alexander Warg. Taming subsystems: Capabilities as universal resource access control in l4. In *Proceedings of the Second Workshop on Isolation and Integration in Embedded Systems*, IIES '09, pages 25–30, New York, NY, USA, 2009. ACM.

[241] Renaud De Landtsheer, Jean christophe Deprez, and Christophe Ponsard. Optimal mapping of task-based computation models over heterogeneous hardware using placer. In *Proc. MODELS 18 Companion (October 2018)*. https://doi.org/https://doi.org/10.1145/3270112.3270136.

[242] Yann LeCun and Yoshua Bengio. The handbook of brain theory and neural networks. chapter Convolutional Networks for Images, Speech,

and Time Series, pages 255–258. MIT Press, Cambridge, MA, USA, 1998.

[243] Jaekyu Lee, Hyesoon Kim, and Richard Vuduc. When prefetching works, when it doesn’t, and why. *ACM Trans. Archit. Code Optim.*, 9(1):2:1–2:29, March 2012.

[244] S. Lee and J. S. Vetter. Early evaluation of directive-based gpu programming models for productive exascale computing. In *SC '12: Proceedings of the International Conference on High Performance Computing, Networking, Storage and Analysis*, pages 1–11, Nov 2012.

[245] Seyong Lee, Seung-Jai Min, and Rudolf Eigenmann. Openmp to gpgpu: a compiler framework for automatic translation and optimization. *ACM Sigplan Notices*, 44(4):101–110, 2009.

[246] Legato Project. LEGaTO Homepage, 2018.

[247] Charles E. Leiserson. Fat-trees: Universal networks for hardware-efficient supercomputing. *IEEE Trans. Comput.*, 34(10):892–901, 1985.

[248] Juri Lelli, Claudio Scordino, Luca Abeni, and Dario Faggioli. Deadline scheduling in the linux kernel. *Software: Practice and Experience*, 46(6):821–839, 16.

[249] KEVIN Lepak, GERRY Talbot, SEAN White, NOAH Beck, S Naffziger, SENIOR FELLOW, et al. The next generation amd enterprise server product architecture. In *Proc. Hot Chips*, pages 1–22, 2017.

[250] Peilong Li, Yan Luo, Ning Zhang, and Yu Cao. Heterospark: A heterogeneous cpu/gpu spark platform for machine learning algorithms. In *2015 IEEE International Conference on Networking, Architecture and Storage (NAS)*, pages 347–348, Aug 2015.

[251] Chunhua Liao, Daniel J. Quinlan, Thomas Panas, and Bronis R. de Supinski. A rose-based openmp 3.0 research compiler supporting multiple runtime libraries. In Mitsuhisa Sato, Toshihiro Hanawa, Matthias S. Müller, Barbara M. Chapman, and Bronis R. de Supinski, editors, *Beyond Loop Level Parallelism in OpenMP: Accelerators, Tasking and More*, pages 15–28, Berlin, Heidelberg, 2010. Springer Berlin Heidelberg.

[252] Jochen Liedtke. Improving ipc by kernel design. *ACM SIGOPS operating systems review*, 27(5):175–188, 1993.

[253] G. Lipari and S. Baruah. Greedy reclamation of unused bandwidth in constant-bandwidth servers. In *Proceedings 12th Euromicro Conference on Real-Time Systems. Euromicro RTS 2000*, pages 193–200, June 2000.

[254] C. L. Liu and James W. Layland. Scheduling algorithms for multiprogramming in a hard-real-time environment. *J. ACM*, 20:46–61, 1973.

[255] Gabriel H. Loh, Yuan Xie, and Bryan Black. Processor design in 3d die-stacking technologies. *Micro, IEEE*, 27(3):31–48, May 2007.

[256] Fan Long and Martin Rinard. Automatic patch generation by learning correct code. In *ACM SIGPLAN Notices*, volume 51, pages 298–312. ACM, 2016.

[257] Francesc Lordan, Enric Tejedor, Jorge Ejarque, Roger Rafanell, Javier Alvarez, Fabrizio Marozzo, Daniele Lezzi, Raül Sirvent, Domenico Talia, and Rosa M Badia. Servicess: an interoperable programming framework for the cloud. *Journal of Grid Computing*, 12(1):67–91, 3 2014.

[258] ARM Ltd. Neon, 2015. http://www.arm.com/products/processors/technologies/neon.php.

[259] Chi-Keung Luk, Sunpyo Hong, and Hyesoon Kim. Qilin: Exploiting parallelism on heterogeneous multiprocessors with adaptive mapping. *2009 42nd Annual IEEE/ACM International Symposium on Microarchitecture (MICRO)*, pages 45–55, 2009.

[260] M. C. Silva Filho et al. Cloudsim plus: A modern java 8 framework for modelling and simulation of cloud computing infrastructures and services.

[261] Ravi Mahajan, Robert Sankman, Neha Patel, Dae-Woo Kim, Kemal Aygun, Zhiguo Qian, Yidnekachew Mekonnen, Islam Salama, Sujit Sharan, Deepti Iyengar, et al. Embedded multi-die interconnect bridge (emib)– a high density, high bandwidth packaging interconnect. In *2016 IEEE 66th Electronic Components and Technology Conference (ECTC)*, pages 557–565. IEEE, 2016.

[262] Redowan Mahmud, Ramamohanarao Kotagiri, and Rajkumar Buyya. Fog computing: A taxonomy, survey and future directions. In *Internet of everything*, pages 103–130. Springer, 2018.

[263] Pavlos Malakonakis, Konstantinos Georgopoulos, Aggelos Ioannou, Luciano Lavagno, Ioannis Papaefstathiou, and Iakovos Mavroidis. Hls algorithmic explorations for hpc execution on reconfigurable hardware— ecoscale. In Nikolaos Voros, Michael Huebner, Georgios Keramidas, Diana Goehringer, Christos Antonopoulos, and Pedro C. Diniz, editors, *Applied Reconfigurable Computing. Architectures, Tools, and Applications*, pages 724–736, Cham, 2018. Springer International Publishing.

[264] James Malcolm, Pavan Yalamanchili, Chris McClanahan, Vishwanath Venugopalakrishnan, Krunal Patel, and John Melonakos. ArrayFire: a GPU acceleration platform. In *Modeling and Simulation for Defense Systems and Applications VII*, volume 8403, 2012.

[265] Grzegorz Malewicz, Matthew H Austern, Aart JC Bik, James C Dehnert, Ilan Horn, Naty Leiser, and Grzegorz Czajkowski. Pregel: a system for large-scale graph processing. In *Proceedings of the 2010 ACM SIGMOD International Conference on Management of data*, pages 135–146. ACM, 2010.

[266] R. Mancuso, R. Dudko, E. Betti, M. Cesati, M. Caccamo, and R. Pellizzoni. Real-time cache management framework for multi-core architectures. In *2013 IEEE 19th Real-Time and Embedded Technology and Applications Symposium (RTAS)*, pages 45–54, April 2013.

[267] Marinos Tsantekidis and Vassilis Prevelakis. Library-Level Policy Enforcement. In *SECURWARE: The Eleventh International Conference on Emerging Security Information, Systems and Technologies*, 2017.

[268] Stefano Markidis, Steven Wei Der Chien, Erwin Laure, Ivy Bo Peng, and Jeffrey S Vetter. Nvidia tensor core programmability, performance & precision. In *2018 IEEE International Parallel and Distributed Processing Symposium Workshops (IPDPSW)*, pages 522–531. IEEE, 2018.

[269] Iakovos Mavroidis, Ioannis Papaefstathiou, Luciano Lavagno, Dimitrios S. Nikolopoulos, Dirk Koch, John Goodacre, Ioannis Sourdis, Vassilis Papaefstathiou, Marcello Coppola, and Manuel Palomino. ECOSCALE: reconfigurable computing and runtime system for future exascale systems. In *2016 Design, Automation & Test in Europe Conference & Exhibition, DATE 2016, Dresden, Germany, March 14-18, 2016*, pages 696–701, 2016.

[270] Maxeler, Inc. The open spatial programming language, 2014. https://openspl.org.

[271] Somnath Mazumdar and Marco Pranzo. Power efficient server consolidation for cloud data center. *Future Generation Computer Systems*, 70:4–16, 2017.

[272] X. Mei, X. Chu, H. Liu, Y. W. Leung, and Z. Li. Energy efficient real-time task scheduling on cpu-gpu hybrid clusters. In *IEEE INFOCOM 2017 - IEEE Conference on Computer Communications*, pages 1–9, May 2017.

[273] Peter M. Mell and Timothy Grance. Sp 800-145. the nist definition of cloud computing. Technical report, Gaithersburg, MD, United States, 2011.

[274] Mark Meredith and Bhuvan Urgaonkar. On Exploiting Resource Diversity in the Public Cloud for Modeling Application Performance. In *CLOUD COMPUTING 2017, The Eighth International Conference on Cloud Computing, GRIDs, and Virtualization*, pages 66–72, Athens, Greece, 2017. IARIA.

[275] Rashid Mijumbi, Joan Serrat, Juan-Luis Gorricho, Niels Bouten, Filip De Turck, and Raouf Boutaba. Network function virtualization: State-of-the-art and research challenges. *IEEE Communications Surveys & Tutorials*, 18(1):236–262, 2016.

[276] Sparsh Mittal and Jeffrey S Vetter. A survey of cpu-gpu heterogeneous computing techniques. *ACM Computing Surveys (CSUR)*, 47(4):69, 2015.

[277] A. K. Mok. Fundamental design problems of distributed systems for the hard-real-time environment. Technical report, Massachussets University of Technology, Cambridge, MA, USA, 1983.

[278] Aloysius K. Mok and Deji Chen. A multiframe model for real-time tasks. In *RTSS*, 1996.

[279] Aaftab Munshi, Benedict Gaster, Timothy G Mattson, and Dan Ginsburg. *OpenCL programming guide*. Pearson Education, 2011.

[280] Zainalabedin Navabi. *VHDL: Analysis and modeling of digital systems*. McGraw-Hill, Inc., 1997.

[281] Stephen Neuendorffer and Fernando Martinez-Vallina. Building Zynq®Accelerators with Vivado®High Level Synthesis. In *Proceedings of the ACM/SIGDA International Symposium on Field Programmable Gate Arrays*, FPGA '13, pages 1–2, New York, NY, USA, 2013. ACM.

[282] Bill Nitzberg and Virginia Lo. Distributed shared memory: A survey of issues and algorithms. *Distributed Shared Memory-Concepts and Systems*, pages 42–50, 1991.

[283] Jose Nunez-Yanez. Energy efficient reconfigurable computing with adaptive voltage and logic scaling. *SIGARCH Comput. Archit. News*, 42(4):87–92, December 2014.

[284] Eriko Nurvitadhi, Jeffrey Cook, Asit Mishra, Debbie Marr, Kevin Nealis, Philip Colangelo, Andrew Ling, Davor Capalija, Utku Aydonat, Aravind Dasu, et al. In-package domain-specific asics for intel® stratix® 10 fpgas: A case study of accelerating deep learning using tensortile asic. In *2018 28th International Conference on Field Programmable Logic and Applications (FPL)*, pages 106–1064. IEEE, 2018.

[285] NVCaffe. Project Website: https://docs.nvidia.com/deeplearning/dgx/caffe-user-guide/index.html.

[286] NVIDIA. NVIDIA GRID Remote Workstation Certifications.

[287] NVIDIA. The Tegra X1 Platform, 2015.

[288] Nvidia. Cuda c programming guide v8.0. 2016.

[289] NVIDIA. NVIDIA DRIVE PX: scalable AI Supercomputer For Autonomous Driving, 2017.

[290] CUDA Nvidia. Compute unified device architecture programming guide. 2007.

[291] S.R. Ohring, M. Ibel, S.K. Das, and M.J. Kumar. On generalized fat trees. In *Parallel Processing Symposium, 1995. Proceedings., 9th International*, pages 37–44, Apr 1995.

[292] Tom Oinn, Matthew Addis, Justin Ferris, Darren Marvin, Martin Senger, Mark Greenwood, Tim Carver, Kevin Glover, Matthew R Pocock, Anil Wipat, et al. Taverna: a tool for the composition and enactment of bioinformatics workflows. *Bioinformatics*, 20(17):3045–3054, 2004.

[293] OpenACC-Standard.org. The OpenACC Application Programming Interface 1.0, November 2011.

[294] OpenCL, 2009. http://www.khronos.org/opencl/.

[295] OpenMP Architecture Review Board. OpenMP application program interface version 4.5, May 2011.

[296] OpenMP Architecture Review Board. OpenMP application program interface version 4.0, July 2013.

[297] Diego Oriato, Stephen Girdlestone, and Oskar Mencer. Chapter three - dataflow computing in extreme performance conditions. In Ali R. Hurson and Veljko Milutinovic, editors, *Dataflow Processing*, volume 96 of *Advances in Computers*, pages 105 – 137. Elsevier, 2015.

[298] Gabriel Ortiz, Lars Svensson, Erik Alveflo, and Per Larsson-Edefors. Instruction level energy model for the adapteva epiphany multi-core processor. In *Proceedings of the Computing Frontiers Conference*, pages 380–384. ACM, 2017.

[299] OscaR Team. OscaR: Operational research in Scala, 2012. Available under the LGPL licence from bitbucket.org/oscarlib/oscar.

[300] J.D. Owens, W.J. Dally, R. Ho, D.N. Jayasimha, S.W. Keckler, and Li-Shiuan Peh. Research challenges for on-chip interconnection networks. *Micro, IEEE*, 27:96–108, Sept 2007.

[301] E. Papadogiannaki, L. Koromilas, G. Vasiliadis, and S. Ioannidis. Efficient Software Packet Processing on Heterogeneous and Asymmetric Hardware Architectures. *IEEE/ACM Transactions on Networking*, 25(3):1593–1606, June 2017.

[302] Antonis Papadogiannakis, Laertis Loutsis, Vassilis Papaefstathiou, and Sotiris Ioannidis. ASIST: Architectural Support for Instruction Set Randomization. In *ACM Conference on Computer and Communications Security (CCS)*, pages 981–992, 2013.

[303] N. Papakonstantinou, F. S. Zakkak, and P. Pratikakis. Hierarchical parallel dynamic dependence analysis for recursively task-parallel programs. In *2016 IEEE International Parallel and Distributed Processing Symposium (IPDPS)*, pages 933–942, May 2016.

[304] Mark S Papamarcos and Janak H Patel. A low-overhead coherence solution for multiprocessors with private cache memories. In *ACM SIGARCH Computer Architecture News*, volume 12, pages 348–354. ACM, 1984.

[305] Yoonho Park, Eric Van Hensbergen, Marius Hillenbrand, Todd Inglett, Bryan Rosenburg, Kyung Dong Ryu, and Robert W Wisniewski. Fusedos: Fusing lwk performance with fwk functionality in a heterogeneous environment. In *Computer Architecture and High Performance Computing (SBAC-PAD), 2012 IEEE 24th International Symposium on*, pages 211–218. IEEE, 2012.

[306] The AUTOSAR partnership. Automotive open system architecture (autosar).

[307] J Thomas Pawlowski. Hybrid memory cube (hmc). In *2011 IEEE Hot chips 23 symposium (HCS)*, pages 1–24. IEEE, 2011.

[308] M. Pelcat, K. Desnos, J. Heulot, C. Guy, J.-F. Nezan, and S. Aridhi. Preesm: A dataflow-based rapid prototyping framework for simplifying multicore dsp programming. In *Education and Research Conference (EDERC), 2014 6th European Embedded Design in*, pages 36–40, Sept 2014.

[309] Rodolfo Pellizzoni, Emiliano Betti, Stanley Bak, Gang Yao, John Criswell, Marco Caccamo, and Russell Kegley. A predictable execution model for cots-based embedded systems. In *2011 17th IEEE Real-Time and Embedded Technology and Applications Symposium*, pages 269–279. IEEE, 2011.

[310] V. Petrucci, M. A. Laurenzano, J. Doherty, Y. Zhang, D. Mossé, J. Mars, and L. Tang. Octopus-man: Qos-driven task management for heterogeneous multicores in warehouse-scale computers. In *2015 IEEE 21st International Symposium on High Performance Computer Architecture (HPCA)*, pages 246–258, Feb 2015.

[311] Khoa Dang Pham, Edson L. Horta, and Dirk Koch. BITMAN: A tool and API for FPGA bitstream manipulations. In *Design, Automation & Test in Europe Conference & Exhibition, DATE 2017, Lausanne, Switzerland, March 27-31, 2017*, pages 894–897, 2017.

[312] Christian Pilato and Fabrizio Ferrandi. Bambu: A free framework for the high-level synthesis of complex applications. 10 2018.

[313] L. Pilosu, L. Mossucca, A. Scionti, S. Ciccia, and et al. Low Power Computing and Communication System for Critical Environments. In *Proc. of the 11th International Conference on P2P, Parallel, Grid, Cloud and Internet Computing (3PGCIC), Asan, Korea*, pages 221–232, Oct. 2016.

[314] Luís Miguel Pinho, Vincent Nélis, Patrick Meumeu Yomsi, Eduardo Quiñones, Marko Bertogna, Paolo Burgio, Andrea Marongiu, Claudio Scordino, Paolo Gai, Michele Ramponi, and Michal Mardiak. P-SOCRATES: A parallel software framework for time-critical many-core systems. *Microprocessors and Microsystems—Embedded Hardware Design*, 39(8):1190–1203, 2015.

[315] The Portland Group. *PGI Accelerator Programming Model for Fortran & C*, 2010.

[316] Georgios Portokalidis and Angelos D. Keromytis. Fast and Practical Instruction-Set Randomization for Commodity Systems. In *Annual Computer Security Applications Conference (ACSAC)*, 2010.

[317] Cédric Pralet. An incomplete constraint-based system for scheduling with renewable resources. In J. Christopher Beck, editor, *Principles*

and Practice of Constraint Programming, pages 243–261, Cham, 2017. Springer International Publishing.

[318] Preesm team.   Preesm rapid prototyping tool, 2018.   `preesm. insa-rennes.fr/website`.

[319] Vassilis Prevelakis and Mohammad Hamad. Extending the operational envelope of applications. In *8th International Conference on Trust & Trustworthy Computing (TRUST 2015)*, 2015.

[320] Jelica Protic, Milo Tomasevic, and Veljko Milutinović. *Distributed shared memory: Concepts and systems*, volume 21. John Wiley & Sons, 1998.

[321] J Ross Quinlan. *C4. 5: programs for machine learning*. Elsevier, 2014.

[322] Steve Ragan. DDoS knocks down DNS, data centers across the U.S. affected.   https://www.csoonline.com/article/3133992/security/ddos-knocks-down-dns-datacenters-across-the-u-s-affected.html,   October 2016.

[323] Naveen Rao. Intel® nervana™ neural network processors (nnp) redefine ai silicon. *Intel https://ai. intel. com/intel-nervana-neural-network-processors-nnp-redefine-ai-silicon*, 2018.

[324] Rapita Systems Corp. *RapiTime, http://www.rapitasystems.com*.

[325] Renaud De Landtsheer. Placer tool, 2018. Available under the LGPL licence from `github.com/TANGO-Project/placer`.

[326] Renaud De Landtsheer. Reference on placer input format. Technical report, CETIC, 2018. Delivered with the releases of the Placer tool [325].

[327] Jordan Robertson and Michael Riley.   The Big Hack: How China Used a Tiny Chip to Infiltrate U.S. Companies. https://www.bloomberg.com/news/features/2018-10-04/the-big-hack-how-china-used-a-tiny-chip-to-infiltrate-america-s-top-companies, October 2018.

[328] Matthew Rocklin. Dask: Parallel computation with blocked algorithms and task scheduling. In *Proceedings of the 14th Python in Science Conference*, pages 130–136. Citeseer, 2015.

[329] P. Ruiu, A. Scionti, J. Nider, and M. Rapoport. Workload management for power efficiency in heterogeneous data centers. In *2016 10th International Conference on Complex, Intelligent, and Software Intensive Systems (CISIS)*, pages 23–30, July 2016.

[330] Florentino Sainz and et al. Leveraging ompss to exploit hardware accelerators. In *26th IEEE International Symposium on Computer Architecture and High Performance Computing, SBAC-PAD 2014, Paris, France, October 22-24, 2014*, pages 112–119, 2014.

[331] Mazeiar Salehie and Ladan Tahvildari. Self-adaptive software: Landscape and research challenges. *ACM Transactions on Autonomous and Adaptive Systems (TAAS)*, 4(2):14, 2009.

[332] Erik Saule, Kamer Kaya, and Ümit V Çatalyürek. Performance evaluation of sparse matrix multiplication kernels on intel xeon phi. In *Parallel Processing and Applied Mathematics*, pages 559–570. Springer, 2013.

[333] Johannes Schemmel, Daniel Briiderle, Andreas Griibl, Matthias Hock, Karlheinz Meier, and Sebastian Millner. A wafer-scale neuromorphic hardware system for large-scale neural modeling. In *Proceedings of 2010 IEEE International Symposium on Circuits and Systems*, pages 1947–1950. IEEE, 2010.

[334] Moritz Schmid, Christian Schmitt, Frank Hannig, Gorker Alp Malazgirt, Nehir Sonmez, Arda Yurdakul, and Adrian Cristal. Big data and hpc acceleration with vivado hls. In *FPGAs for Software Programmers*, pages 115–136. Springer, 2016.

[335] Jürgen Schmidhuber. Deep learning in neural networks: An overview. *Neural networks : the official journal of the International Neural Network Society*, 61:85–117, 2015.

[336] Alberto Scionti, Somnath Mazumdar, and Antoni Portero. Efficient data-driven task allocation for future many-cluster on-chip systems. In *2017 International Conference on High Performance Computing & Simulation (HPCS)*, pages 503–510. IEEE, 2017.

[337] Alberto Scionti, Somnath Mazumdar, and Stéphane Zuckerman. Enabling massive multi-threading with fast hashing. *IEEE Computer Architecture Letters*, 17(1):1–4, 2018.

[338] Alberto Scionti, Pietro Ruiu, Olivier Terzo, Joel Nider, Craig Petrie, and Niccolo Baldoni. Opera: A low power approach to the next generation cloud infrastructures. In *Digital System Design (DSD), 2016 Euromicro Conference on*, pages 326–333. IEEE, 2016.

[339] Oren Segal, Philip Colangelo, Nasibeh Nasiri, Zhuo Qian, and Martin Margala. Sparkcl: A unified programming framework for accelerators on heterogeneous clusters. *CoRR*, abs/1505.01120, 2015.

[340] M. A. Serrano, A. Melani, R. Vargas, A. Marongiu, M. Bertogna, and E. Quiñones. Timing characterization of openmp4 tasking model. In *2015 International Conference on Compilers, Architecture and Synthesis for Embedded Systems (CASES)*, pages 157–166, Oct 2015.

[341] V. Shankar and S. Chang. Performance of caffe on qct deep learning reference architecture – a preliminary case study. In *2017 IEEE 4th International Conference on Cyber Security and Cloud Computing (CSCloud)*, pages 35–39, June 2017.

[342] H. Sharma, J. Park, D. Mahajan, E. Amaro, J. K. Kim, C. Shao, A. Mishra, and H. Esmaeilzadeh. From high-level deep neural models

to fpgas. In *2016 49th Annual IEEE/ACM International Symposium on Microarchitecture (MICRO)*, pages 1–12, Oct 2016.

[343] Weisong Shi, Jie Cao, Quan Zhang, Youhuizi Li, and Lanyu Xu. Edge computing: Vision and challenges. *IEEE Internet of Things Journal*, 3:637–646, 2016.

[344] Siemens. The Jailhouse Hypervisor.

[345] C Silvano, G Agosta, J Barbosa, A Bartolini, A R Beccari, L Benini, J Bispo, J M P Cardoso, C Cavazzoni, S Cherubin, R Cmar, D Gadioli, C Manelfi, J Martinovič, R Nobre, G Palermo, M Palkovič, P Pinto, E Rohou, N Sanna, and K Slaninová. The ANTAREX tool flow for monitoring and autotuning energy efficient HPC systems. In *2017 International Conference on Embedded Computer Systems: Architectures, Modeling, and Simulation (SAMOS)*, pages 308–316, 2017.

[346] C Silvano, G Agosta, A Bartolini, A R Beccari, L Benini, J Bispo, R Cmar, J M P Cardoso, C Cavazzoni, J Martinovič, G Palermo, M Palkovič, P Pinto, E Rohou, N Sanna, and K Slaninová. Autotuning and adaptivity approach for energy efficient Exascale HPC systems: The ANTAREX approach. In *2016 Design, Automation & Test in Europe Conference & Exhibition (DATE)*, pages 708–713, 2016.

[347] Avinash Sodani. Knights landing (knl): 2nd generation intel® xeon phi processor. In *2015 IEEE Hot Chips 27 Symposium (HCS)*, pages 1–24. IEEE, 2015.

[348] OASIS Standard. Topology and orchestration specification for cloud applications version 1.0, 2013.

[349] Martin Stigge, Pontus Ekberg, Nan Guan, and Wang Yi. The digraph real-time task model. *2011 17th IEEE Real-Time and Embedded Technology and Applications Symposium*, pages 71–80, 2011.

[350] Martin Stigge, Pontus Ekberg, Nan Guan, and Wang Yi. On the tractability of digraph-based task models. *2011 23rd Euromicro Conference on Real-Time Systems*, pages 162–171, 2011.

[351] John E Stone, David Gohara, and Guochun Shi. Opencl: A parallel programming standard for heterogeneous computing systems. *Computing in science & engineering*, 12(1-3):66–73, 2010.

[352] D Stroobandt, A L Varbanescu, C B Ciobanu, M Al Kadi, A Brokalakis, G Charitopoulos, T Todman, X Niu, D Pnevmatikatos, A Kulkarni, E Vansteenkiste, W Luk, M D Santambrogio, D Sciuto, M Huebner, T Becker, G Gaydadjiev, A Nikitakis, and A J W Thom. EXTRA: Towards the exploitation of eXascale technology for reconfigurable architectures. In *2016 11th International Symposium on Reconfigurable Communication-centric Systems-on-Chip (ReCoSoC)*, pages 1–7, jun 2016.

[353] Y. Sukuki, S. Kato, H. Yamada, and K. Kono. GPUvm: why not virtualizing GPUs at the hypervisor? In *2014 USENIC Annual Technical Conference*, 2014.

[354] Yifan Sun, Saoni Mukherjee, Trinayan Baruah, Shi Dong, Julian Gutierrez, Prannoy Mohan, and David Kaeli. Evaluating performance tradeoffs on the radeon open compute platform. In *2018 IEEE International Symposium on Performance Analysis of Systems and Software (ISPASS)*, pages 209–218. IEEE, 2018.

[355] CK Tang. Cache system design in the tightly coupled multiprocessor system. In *Proceedings of the June 7-10, 1976, national computer conference and exposition*, pages 749–753. ACM, 1976.

[356] TANGO Team. Tango: Simplify and optimize heterogeneity, 2018. `tango-project.eu`.

[357] Obeo team. Eclipse sirius, 2018. `https://www.obeo.fr/en/products/eclipse-sirius`.

[358] Mellanox Technologies. Mellanox bluefield® smartnic – high performance ethernet network adapter cards. `http://www.mellanox.com/related-docs/prod_adapter_cards/PB_BlueField_Smart_NIC.pdf`.

[359] Mellanox Technologies. Infiniband in the enterprise data center. In *White Paper*, 2006.

[360] Xavier Teruel, Xavier Martorell, Alejandro Duran, Roger Ferrer, and Eduard Ayguadé. Support for openmp tasks in nanos v4. In *Proceedings of the 2007 conference of the center for advanced studies on Collaborative research*, pages 256–259. IBM Corp., 2007.

[361] Texas Instruments. Multicore dsp+arm keystone ii system-on-chip (soc), 2016.

[362] Douglas Thain, Todd Tannenbaum, and Miron Livny. Distributed computing in practice: the condor experience. *Concurrency and computation: practice and experience*, 17(2-4):323–356, 2005.

[363] The Hercules Consortium. Deliverable D5.3 – Integrated schedulability analysis. Technical report, The Hercules Consortium, 2018.

[364] The P-SOCRATES consortium. P-SOCRATES – Parallel Software Framework for Time-Critical Many-core Systems, 2015.

[365] Donald E Thomas and Philip R Moorby. *The Verilog® Hardware Description Language*, volume 2. Springer Science & Business Media, 2002.

[366] TOP 500. TOP 500, June 2017, 2017. [Online; accessed 5-June-2017].

[367] Abderezak Touzene and Khaled Day. All-to-all broadcasting in torus network on chip. *The Journal of Supercomputing*, 71(7), 2015.

[368] Theodore Tryfonas, Dimitris Gritzalis, and Spyros Kokolakis. A qualitative approach to information availability. In *IFIP International Information Security Conference*, pages 37–47. Springer, 2000.

[369] Alexey Tumanov, Timothy Zhu, Jun Woo Park, Michael A Kozuch, Mor Harchol-Balter, and Gregory R Ganger. Tetrisched: global rescheduling with adaptive plan-ahead in dynamic heterogeneous clusters. In *Proceedings of the Eleventh European Conference on Computer Systems*, page 35. ACM, 2016.

[370] George Tzenakis, Angelos Papatriantafyllou, Hans Vandierendonck, Polyvios Pratikakis, and Dimitrios S Nikolopoulos. Bddt: Block-level dynamic dependence analysis for task-based parallelism. In *International Workshop on Advanced Parallel Processing Technologies*, pages 17–31. Springer, 2013.

[371] Giorgos Vasiliadis, Spiros Antonatos, Michalis Polychronakis, Evangelos P. Markatos, and Sotiris Ioannidis. Gnort: High Performance Network Intrusion Detection Using Graphics Processors. In *Proceedings of the 11th International Symposium on Recent Advances in Intrusion Detection*, RAID '08, pages 116–134, Berlin, Heidelberg, 2008. Springer-Verlag.

[372] Vinod Kumar Vavilapalli, Arun C. Murthy, Chris Douglas, Sharad Agarwal, Mahadev Konar, Robert Evans, Thomas Graves, Jason Lowe, Hitesh Shah, Siddharth Seth, Bikas Saha, Carlo Curino, Owen O'Malley, Sanjay Radia, Benjamin Reed, and Eric Baldeschwieler. Apache hadoop yarn: Yet another resource negotiator. In *Proceedings of the 4th Annual Symposium on Cloud Computing*, SOCC '13, pages 5:1–5:16, New York, NY, USA, 2013. ACM.

[373] S. Vestal. Preemptive scheduling of multi-criticality systems with varying degrees of execution time assurance. In *28th IEEE International Real-Time Systems Symposium (RTSS 2007)*, pages 239–243, Dec 2007.

[374] Vineyard.   Objectives and rationales of the project.   [online],   2018.   {"http://vineyard-h2020.eu/en/project/objectives-and-rationale-of-the-project.html"}.

[375] Mattias De Wael, Stefan Marr, Bruno De Fraine, Tom Van Cutsem, and Wolfgang De Meuter. Partitioned global address space languages. *ACM Computing Surveys (CSUR)*, 47(4):62, 2015.

[376] Pete Warden. Why gemm is at the heart of deep learning. 20 April 2015.

[377] Yuan Wen and Michael FP O'Boyle. Merge or separate?: Multi-job scheduling for opencl kernels on cpu/gpu platforms. In *Proceedings of the General Purpose GPUs*, pages 22–31. ACM, 2017.

[378] Yuan Wen, Zheng Wang, and Michael FP O'boyle. Smart multi-task scheduling for opencl programs on cpu/gpu heterogeneous platforms. In *High Performance Computing (HiPC), 2014 21st International Conference on*, pages 1–10. IEEE, 2014.

[379] David Wentzlaff and Anant Agarwal. Factored operating systems (fos): The case for a scalable operating system for multicores. *SIGOPS Oper. Syst. Rev.*, 43(2):76–85, April 2009.

[380] Wikipedia. SCHED_DEADLINE.

[381] Michael Wilde, Mihael Hategan, Justin M Wozniak, Ben Clifford, Daniel S Katz, and Ian Foster. Swift: A language for distributed parallel scripting. *Parallel Computing*, 37(9):633–652, 2011.

[382] Robert W Wisniewski, Todd Inglett, Pardo Keppel, Ravi Murty, and Rolf Riesen. mos: An architecture for extreme-scale operating systems. In *Proceedings of the 4th International Workshop on Runtime and Operating Systems for Supercomputers*, page 2. ACM, 2014.

[383] Xilinx, Inc., . The Xlinx Ultrascale Architecture.

[384] Xilinx, Inc. Vivado High-Level Synthesis, September 2017. http://www.xilinx.com/hls.

[385] Xilinx, Inc. Sdsoc development environment. [online], July 2018. https://www.xilinx.com/sdsoc.

[386] Yonghong Yan, Max Grossman, and Vivek Sarkar. Jcuda: A programmer-friendly interface for accelerating java programs with cuda. In Henk Sips, Dick Epema, and Hai-Xiang Lin, editors, *Euro-Par 2009 Parallel Processing*, pages 887–899, Berlin, Heidelberg, 2009. Springer Berlin Heidelberg.

[387] Chao-Tung Yang, Chih-Lin Huang, and Cheng-Fang Lin. Hybrid cuda, openmp, and mpi parallel programming on multicore gpu clusters. *Computer Physics Communications*, 182(1):266–269, 2011.

[388] G. Yao, H. Yun, Z. P. Wu, R. Pellizzoni, M. Caccamo, and L. Sha. Schedulability analysis for memory bandwidth regulated multicore real-time systems. *IEEE Transactions on Computers*, 65(2):601–614, Feb 2016.

[389] Y. Yuan, M. F. Salmi, Y. Huai, K. Wang, R. Lee, and X. Zhang. Spark-gpu: An accelerated in-memory data processing engine on clusters. In *2016 IEEE International Conference on Big Data (Big Data)*, pages 273–283, Dec 2016.

[390] Herve Yviquel, Antoine Lorence, Khaled Jerbi, Gildas Cocherel, Alexandre Sanchez, and Mickael Raulet. Orcc: Multimedia development made easy. In *Proceedings of the 21st ACM International Conference on Multimedia*, MM '13, pages 863–866, New York, NY, USA, 2013. ACM.

[391] Nikolaos Zacheilas, Nikolaos Chalvantzis, Ioannis Konstantinou, Vana Kalogeraki, and Nectarios Koziris. Orion: Online resource negotiator for multiple big data analytics frameworks. In *Autonomic Computing (ICAC), 2018 IEEE International Conference on*. IEEE, 2018.

[392] Matei Zaharia, Mosharaf Chowdhury, Michael J Franklin, Scott Shenker, and Ion Stoica. Spark: Cluster computing with working sets. *HotCloud*, 10(10-10):95, 2010.

[393] Matei Zaharia, Reynold S. Xin, Patrick Wendell, Tathagata Das, Michael Armbrust, Ankur Dave, Xiangrui Meng, Josh Rosen, Shivaram Venkataraman, Michael J. Franklin, Ali Ghodsi, Joseph Gonzalez, Scott Shenker, and Ion Stoica. Apache spark: A unified engine for big data processing. *Commun. ACM*, 59(11):56–65, October 2016.

[394] F. S. Zakkak, D. Chasapis, P. Pratikakis, A. Bilas, and D. S. Nikolopoulos. Inference and declaration of independence: Impact on deterministic task parallelism. In *2012 21st International Conference on Parallel Architectures and Compilation Techniques (PACT)*, pages 453–454, Sept 2012.

[395] Foivos S. Zakkak, Dimitrios Chasapis, Polyvios Pratikakis, Angelos Bilas, and Dimitrios S. Nikolopoulos. Inference and declaration of independence in task-parallel programs. In Chenggang Wu and Albert Cohen, editors, *Advanced Parallel Processing Technologies*, pages 1–16, Berlin, Heidelberg, 2013. Springer Berlin Heidelberg.

Index